Practical Applications of Laser Ablation

Edited by Dongfang Yang

Published in London, United Kingdom

IntechOpen

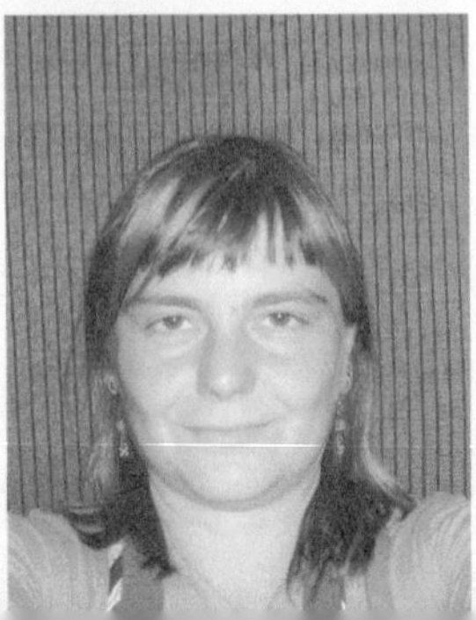

Supporting open minds since 2005

Practical Applications of Laser Ablation
http://dx.doi.org/10.5772/intechopen.92496
Edited by Dongfang Yang

Contributors
Stefan Andrei Irimiciuc, Norina Forna, Andrei Agop, Maricel Agop, Doriana Forna Agop, Stefan Lucian Toma, Jeyaprakash Natarajan, Che-Hua Yang, Bhuvanesh Kumar M, Cyril Robinson Azariah John Chelliah, Rajesh Swaminathan, Jerome Wolfman, Beatrice Negulescu, Ndioba Niang, Antoine Ruyter, Nazir Jaber, Marco A.Z. Arruda, Renata Amais, Danielle S. Da Silva Francischini, Pedro S. Moreau, Raj Kumar Durairaj, Toshio Takiya, Naoaki Fukuda, Saluru Baba Krupanidhi, Deependra Kumar Singh, Karuna Kar Nanda

First published in London, United Kingdom, 2021 by IntechOpen
IntechOpen is the global imprint of INTECHOPEN LIMITED, registered in England and Wales, registration number: 11086078, 5 Princes Gate Court, London, SW7 2QJ, United Kingdom
Printed in Croatia

British Library Cataloguing-in-Publication Data
A catalogue record for this book is available from the British Library

Additional hard and PDF copies can be obtained from orders@intechopen.com

Practical Applications of Laser Ablation
Edited by Dongfang Yang
p. cm.
Print ISBN 978-1-83968-303-9
Online ISBN 978-1-83968-304-6
eBook (PDF) ISBN 978-1-83968-305-3

For EU product safety concerns:
IN TECH d.o.o., Prolaz Marije Krucifikse Kozulić 3, 51000 Rijeka, Croatia, info@intechopen.com or visit our website at intechopen.com.

Meet the editor

Dongfang Yang received his Ph.D. in Physical Chemistry from the University of Guelph in 1995. He joined the National Research Council Canada in London Ontario in 2001 and is now a Senior Research Officer. His current research interests include laser materials processing; pulsed laser, sputtering and e-beam deposition of thin films; new materials development for energy storage devices; chemical and optical sensors development; and electrochemical studies of organic adsorption and self-assembly monolayer. He is currently serving as an editor or editorial board member for ten scientific journals and was listed among the top 2% most-cited scientists according to a Stanford study in 2020. He also holds an adjunct professorship at Western University, Ontario, Canada.

Contents

Preface XI

Section 1
Pulsed Laser Deposition 1

Chapter 1 3
Interface Combinatorial Pulsed Laser Deposition to Enhance Heterostructures Functional Properties
by Jérôme Wolfman, Beatrice Negulescu, Antoine Ruyter, Ndioba Niang and Nazir Jaber

Chapter 2 23
Pulsed Laser Deposition of Transition Metal Dichalcogenides-Based Heterostructures for Efficient Photodetection
by Deependra Kumar Singh, Karuna Kar Nanda and Saluru Baba Krupanidhi

Chapter 3 45
Binary Metal Oxides Thin Films Prepared from Pulsed Laser Deposition
by Cyril Robinson Azariah John Chelliah and Rajesh Swaminathan

Section 2
Laser Surface Modification, Machining and Nanoparticle Formation 59

Chapter 4 61
Laser Surface Modification of Materials
by Natarajan Jeyaprakash, Che-Hua Yang and Durairaj Raj Kumar

Chapter 5 87
Laser Machining
by Natarajan Jeyaprakash, Che-Hua Yang and Manickam Bhuvanesh Kumar

Chapter 6 103
Nanoparticle Formation and Deposition by Pulsed Laser Ablation
by Toshio Takiya and Naoaki Fukuda

Section 3
Plasma Generated by Laser Ablation **119**

Chapter 7 **121**
Dynamics of Transient Plasmas Generated by ns Laser Ablation of Memory Shape Alloys
by Stefan Andrei Irimiciuc, Norina Forna, Andrei Agop, Maricel Agop, Stefan Toma and Doriana Forna Agop

Chapter 8 **137**
Laser Chemical Elemental Analysis: From Total to Images
by Renata S. Amais, Danielle S. Francischini, Pedro S. Moreau and Marco A.Z. Arruda

Preface

This book covers theoretical and experimental aspects of the laser ablation phenomenon for material processing including pulsed laser deposition of thin films, laser surface modification, laser machining and laser nanoparticle formation. It also includes a study of the dynamics of plasmas generated by laser ablation of multicomponent materials and an overview of laser-induced breakdown spectroscopy (LIBS) and laser ablation inductively coupled plasma mass spectrometry (LA-ICP-MS) techniques for chemical analysis. Chapter 1 exploits the use of the Combinatorial Pulsed Laser deposition (CPLD) method to create functional interface libraries of metal oxides by continuously modulating the composition of interfaces at the first few atomic layers to alter their functional properties. Chapter 2 presents an overview of the fabrication and applications of Transition Metal Dichalcogenides (TMDCs)-based photodetectors fabricated by the Pulsed Laser Deposition (PLD) method. It highlights the advantages and limitations of using PLD for fabricating TMDCs-based photodetectors, as well as describes unresolved problems and suggested future directions in this field. Chapter 3 presents the deposition and characterization of thin films of ZnO, MoO_3 and MoO_2 as well as binary oxides such as ZnO/MoO_3 by the PLD method for field-effect transistor applications. Chapter 4 gives a concise overview of various laser-based surface modification processes including laser surface hardening, melting, alloying, cladding and texturing for improving the microstructure, hardness and wear resistance of mechanical components. Chapter 5 presents an excellent overview of laser beam machining techniques. It provides an easy-to-understand explanation of how lasers work, the unique properties of laser beams, types of lasers available, and applications of lasers in the machining of materials. The chapter also describes the construction of laser machining apparatus and the mechanism of material removal. Finally, it discusses the advantages and limitations of laser machining as compared to other conventional machining processes. Chapter 6 gives an overview of using laser ablation in the presence of a background gas to fabricate specific nanoparticles under non-equilibrium states. The chapter describes the mechanism of nanoparticle formation during laser ablation processes and methods to control the inter-particle spacing and patterns of nanoparticles in a film to obtain novel properties. It also describes the application of nanoparticles and thin films deposited for functional devices in several fields. Chapter 7 presents an experimental investigation and theoretical modeling of transient plasmas generated by laser ablation of complex metallic Cu-Mn-Al and Fe-Mn-Si targets. Space- and time-resolved optical emission spectroscopy and fast camera imaging were used to understand the dynamics of multicomponent plasmas. Chapter 8 gives an overview of the LIBS and LA-ICP-MS techniques. It examines the principal, instrumentation, analytical performance, data acquisition and imaging, calibration challenges and sensitivity improvement strategies for both techniques. It also provides representative examples of using LIBS and LA-ICP-MS techniques for chemical analysis in environmental, biological, medical, and forensic research.

Through these eight chapters, which include original research studies and literature reviews written by experts from the international scientific community, the reader will gain a better understanding of the laser ablation process for processing materials and chemical analysis. I gratefully acknowledge all chapter authors for their enthusiastic and collaborative contributions and many thanks to Publishing Process Manager Ms. Mia Vulovic for her guidance and support in the preparation of this book.

Dongfang Yang, Ph.D.
Senior Research Officer,
Automotive and Surface Transportation,
National Research Council Canada,
Ottawa, Ontario, Canada

Section 1

Pulsed Laser Deposition

Chapter 1

Interface Combinatorial Pulsed Laser Deposition to Enhance Heterostructures Functional Properties

Jérôme Wolfman, Beatrice Negulescu, Antoine Ruyter, Ndioba Niang and Nazir Jaber

Abstract

In this chapter we will describe a new development of combinatorial pulsed laser deposition (CPLD) which targets the exploration of interface libraries. The idea is to modulate continuously the composition of interfaces on a few atomic layers in order to alter their functional properties. This unique combinatorial synthesis of interfaces is possible due to very specific PLD characteristics. The first one is its well-known ability for complex oxide stoichiometry transfer from the target to the film. The second one is the layer by layer control of thin film growth at the atomic level using in-situ RHEED characterization. The third one relates to the directionality of the ablated plume which allows for selective area deposition on the substrate using a mobile shadow-mask. However PLD also has some limitations and important PLD aspects to be considered for reliable CPLD are reviewed. Multiple examples regarding the control of interface magnetism in magnetic tunnel junctions and energy band and Schottky barrier height tuning in ferroelectric tunable capacitors are presented.

Keywords: combinatorial synthesis, oxide thin films and multilayers, functional heterostructure

1. Introduction

The first report of laser ablation generated plasma to deposit a film dates back to 1965, using a continuous ruby laser [1]. As the obtained film quality was inferior to those made with other deposition techniques at that time, the laser deposition remained confidential for two decades. The discovery mid 80's of the high Tc superconductor $YBa_2Cu_3O_{7-x}$ (YBCO) [2–3] triggered a frantic search for room temperature superconducting cuprates materials, including strong effort for thin film synthesis. The complex cuprate cationic composition makes film growth difficult using conventional physical vapor deposition (PVD) techniques like magnetron sputtering or evaporation. Indeed, to grow films with the right composition it is mandatory to start with a stoichiometric plasma. The Ar^+ sputtering rate of multi-cationic targets and the distribution in the plasma strongly depends on the cations mass, which precludes the use of a stoichiometric target to obtain a stoichiometric plasma. The offset target composition leading to stoichiometric films is unpredictable and a trial-and-error method is usually employed that renders magnetron sputtering an impractical technique for multi-cationic new materials research. Evaporation temperatures depend on the nature of the cations, which make a thermal vaporization of stoichiometric target inappropriate too. The attempt to produce cuprates films using sputtering and evaporation led to poor quality films.

Ceramic target vaporization based on laser ablation does not depend on the nature of the atoms within the target and result in a stoichiometric plasma as long as the energy per surface unit (i.e. the laser fluence) is above the ablation threshold. Venkatesan and co-workers were the first to vaporize an YBCO target using a pulsed excimer laser (UV KrF). After fine tuning the ablation conditions by checking the spatial distribution of the cations they rapidly produced superconducting films having the best physical properties of that time [4]. This first successful synthesis of high Tc cuprates films using a pulsed laser drew the interest of the community and several laboratories started working in the field now known as pulsed laser deposition (PLD). In the following decade, beside cuprates superconductors, strong interest in colossal magneto-resistive manganites and multiferroic ferrites has arisen. This intense scientific activity on multi-cation oxides associated to PLD ease of use, versatility and reasonable cost contributed to its rapid expansion in the 90's.

Since then PLD based thin film research thrived over a wide range of materials, physical properties and applications. PLD has benefited from in-situ real-time characterization tools developed for other deposition technique to mature into an epitaxial film growth method with control at the atomic scale and able to produce heterostructures with sharp interfaces. At the end of the 90's, material scientists considered transposing combinatorial synthesis, a high throughput synthesis method developed by the pharmaceutical industry, to the field of multi-cation oxides research. Combinatorial PLD (CPLD) rapidly emerged, the basic idea being to grow on a single substrate, in a single deposition run, a film with continuous in-plane chemical modulation. In this way, a wide range of chemical compositions are produced within the same sample and can be quickly scanned to identify compounds with optimum targeted properties. Since the new millennium, CPLD has been refined and its field extended to ternary phase diagram exploration. More recently CPLD research field has been extended to a new territory: the exploration of interface compositions in heterostructures with enhanced functional properties a.k.a. Interface Combinatorial Pulsed Laser Deposition (ICPLD).

2. Pulsed laser deposition for combinatorial synthesis

The aim of this chapter is not to thoroughly describe PLD but to emphasize its most important aspects, advantages and limitations with regard to combinatorial synthesis of oxide films and heterostructures. The reader interested in an exhaustive description of PLD is encouraged to consult PLD's introduction reference book [5] or the other chapters devoted to the subject in this book.

The first pre-requisite for CPLD is the growth of films with uniform thickness and homogeneous composition over the entire sample surface. Although PLD has a reputation for stoichiometric transfer from the target to the film, this is not however straightforward and several deposition parameters have to be fine-tuned. Starting from the formation of a stoichiometric plasma, target thermal vaporization occurring below the ablation threshold should be reduced as much as possible. This means that local fluence everywhere on the laser beam spot should be above the ablation threshold, implying very steep sidewalls of the laser beam energy distribution. Such a distribution, called a top-hat, necessitate laser beam shaping with a beam homogenizer. The laser beam divergence depends on the discharge voltage and could affect the beam spot size and energy distribution depending on the beam shaping method, so the discharge voltage should be kept constant. The plasma expands from the target toward the substrate in the form of a plume which interacts with ambient gas molecules or atoms. Multiple collisions per atom or ions occur leading to a thermalized but still highly directional plasma reaching the substrate. It follows a radial distribution of thickness and composition at the substrate surface. In order to get a uniform and homogeneous film, one has to scan the plume with respect to the substrate, by either scanning the laser spot over the target surface at constant focus and/or moving the substrate (rotation, translation). Doing so will result in a film with much increased thickness uniformity and homogeneous composition. However the composition may not match the target stoichiometry. Indeed, it has been shown that cationic off-stoichiometry up to a few percent occurs depending on the fluence for a fixed spot size and target-substrate distance [6]. So cationic composition/uniformity should always be checked versus fluence with high sensitivity characterization techniques. Once the right fluence has been determined and stabilized over the laser spot on the target, it is crucial to ensure that it stays constant in time. Regarding this point, two factors have to be taken into account. The first one is the degradation of the excimer laser gas charge with time, leading to a drop of the output energy. As already specified, it is advisable to work at constant voltage discharge, so one should not compensate the energy drop with voltage. The solution is to start with a higher energy output than required and to modulate the beam energy with an external attenuator. The second factor is the laser attenuation at the chamber introduction viewport. Although the visible part of the plasma, i.e. the plume, is very forward peaked, some low energetic species are slowly deposited everywhere in the chamber, including on the introduction viewport. This leads to a time varying attenuation which will affect the fluence at the target surface. To compensate for this attenuation, it is mandatory to measure the fluence after the viewport and to increase the input energy accordingly. One can reorient the beam out of the chamber through a pollution preserved port with a translatable UV mirror to measure its energy without breaking the vacuum.

The second pre-requisite for CPLD is a layer by layer growth mode for a control of the composition at the unit cell level. Layer by layer PLD growth of various perovskites has been demonstrated multiple times using in-situ real time grazing incidence electron diffraction (RHEED). This ability for 2d growth comes from the very peculiar and sequential PLD surface crystallization kinetic. Indeed, although average PLD deposition rate is quite slow, instantaneous deposition rate a few μs after laser pulses is extremely high, creating a supersaturation close to the surface which results in high nucleation rate. The large number of nucleus present after the first pulse favors 2d growth. Furthermore, as PLD relies on photon to vaporize the target, the ambient gas pressure can be varied over a very wide range from vacuum level to a few mbar. The deposition pressure allows for the control of the energy of the species reaching the substrate and ultimately their remaining kinetic energy to explore the surface and find the nucleation sites.

The last pre-requisite for CPLD is the ability to lower the average deposition rate per pulse to produce the smallest composition step when mixing materials from different targets. For a fixed fluence and target-substrate distance, the deposition rate per pulse strongly correlates with the laser spot size on the target which is easily adjustable. Deposition rate as low as a few hundred pulses per perovskite unit cell thick layer can be reached. One drawback of PLD is the lack of deposition rate stability. Indeed the structure of the irradiated target surface evolves with laser exposure; the deposition rate lowering with target aging is more severe for a brand new surface and decays with target exposure. Scanning over a large area of the target reduces the number of laser spots per location and makes this problem less stringent. It is however mandatory to calibrate the deposition rate prior a deposition or to measure it in-situ using RHEED.

3. Combinatorial pulsed laser deposition

The concept of combinatorial research, initially introduced in the pharmaceutical industry, aims at synthesizing a large number of compounds with complex and systematically varied composition in a single batch. A high speed characterization technique is then used to scan through this material library and identify the compounds presenting the targeted property.

This approach is very appealing regarding multi-cationic perovskite-related oxides. Indeed their physical properties, covering a very wide span (superconductivity, ferroelectricity, ferromagnetism, colossal magnetoresistance, tunable resistivity ...), result from the subtle equilibrium between competing interactions involving charges, spins and orbitals. This results in an extreme sensitivity of the properties to the cationic composition and requires a fine, thorough and systematic scanning in order to optimize the performances.

The first thin film combinatorial synthesis attempts implied multiple successive room temperature depositions to vary the composition. Several annealing steps were then required for thermal diffusion and crystallization. This approach, where reaction products are dominated by thermodynamic like in ceramic sintering, is not appropriate for epitaxial thin films and heterostructures. On the contrary, epitaxial oxide thin films growth by PLD does not require post-deposition annealing. Moreover the directionality of the plume allows for localized deposition on the substrate through a shadow-mask. These multiple advantages were soon associated to combinatorial synthesis of oxide materials at the end of the 90's simultaneously in the United States and in Japan. The original approach is schematically described in **Figure 1**.

The single unit cell combinatorial cycle consists of two successive depositions from different targets. The material deposited from each target is distributed across the substrate surface using a moving shadow-mask in order to vary locally the layer completeness (from 0–100%). After one deposition cycle, the resulting 1uc thick layer has a composition which varies laterally from one target composition to the other (composition spread). Then this deposition cycle is repeated N times in order to achieve targeted final film thickness.

This idealized combinatorial PLD synthesis produces a continuous variation of compositions on a single sample, guarantying identical growth conditions for all compositions and eliminating the risk of sample variability.

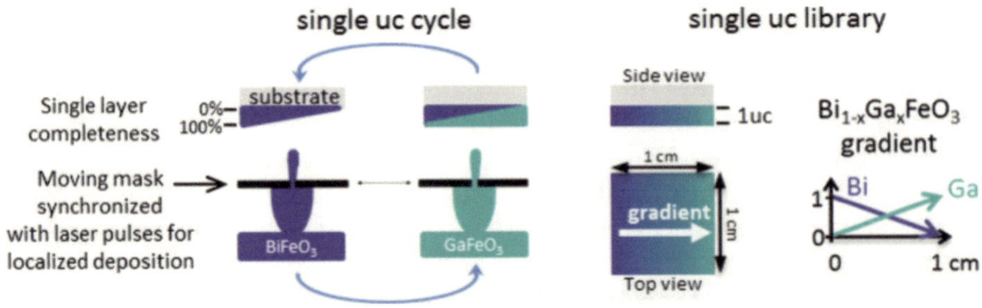

Figure 1.
Schematic of the combinatorial single layer synthesis and of the resulting composition spread library (e.g. $Bi_{1-x}Ga_xFeO_3$ composition library).

A consequence of CPLD is the necessity to use local probes to scan through the compositions library and assess performances. There is a trade-off between the minimum probe size required to measure the targeted physical property and the lateral composition gradient in order to characterize a "homogeneous" compound at the probe level.

The main difficulty regarding CPLD synthesis is to reach adequate control of both local composition and thickness using PLD. As explained in the previous chapter this requires long in-depth preparation work involving chemical characterization and exhaustive optimization of all the deposition parameters.

A prerequisite to reliable CPLD synthesis is to produce a uniform and smooth film with constant composition and thickness from each target over the surface of the future CPLD samples. One should not attempt CPLD before proving this achievement. Another important point is to keep a statistical approach to the characterization of the libraries. It is tempting to produce ternary phase diagrams using 3 different targets on a single substrate. However, in this case there is only one location on the sample per compound which is statistically insufficient. To our knowledge, any published work concerning CPLD ternary phase diagrams on a single sample does not present a proper characterization that demonstrates the control of the composition and thickness across the library. This is not very surprising knowing how difficult it already is to master a single gradient for a binary diagram across one sample. Unfortunately even for binary samples a large number of CPLD articles have been published without any evidence of control of composition/thickness which brought some discredit on CPLD among the scientific community. To change this perception and reinforce CPLD credit, we proposed a decade ago a statistical approach to the characterization of both the composition and the physical properties for binary phase diagram [7]. More recently we developed an alternative approach to explore ternary phase diagrams: instead of trying to produce the full ternary phase diagram on a single sample, we select lines of compositions cutting through the corresponding triangular diagram. Each synthesized sample has a composition gradient in one direction with the compositions defined for one line [8]. This way statistical characterization are possible along the direction orthogonal to the gradient. Scanning through the ternary phase diagram along multiple lines thus requires the synthesis of a few samples with only three targets.

To illustrate the effectiveness of CPLD we will discuss about the search for new lead-free piezoelectrics to replace (1-x)$PbZrO_3$-x$PbTiO_3$ (PZT), the most used material in microelectromechanical systems for sensing, actuating, or energy harvesting applications. PZT ferroelectric films present large piezoelectric coefficients and electromechanical coupling, enabling long range motions and high energy densities [9]. A unique characteristic of lead-based solid solutions presenting high piezoelectric coefficients is the strong enhancement of their piezoelectric response in the vicinity of a composition induced phase transition between ferroelectric phases with different crystalline symmetries, called a morphotropic phase boundary (MPB). In PZT the MPB lies between a rhombohedral ferroelectric phase and a tetragonal ferroelectric phase [10]. The microscopic origin of this enhanced piezoelectric activity is still being debated but usually involves easiness of polarization rotation at the MPBs [11–13]. PZT is lead-based and thus targeted by environmental regulations (e.g. RoHS EU Directive). Thus, alternative piezoelectric lead-free materials are required, and an obvious direction is to look for MPBs in other ferroelectric solid solutions. The rhombohedral perovskite $BiFeO_3$ (BFO), being a robust ferroelectric ($T_c \sim 1100$ K) with record polarization (100 $\mu C{\cdot}cm^{-2}$), is a good starting point [14, 15]. Solid solutions of BFO with tetragonal ferroelectric perovskite phase like $PbTiO_3$ and $BaTiO_3$ have been synthesized, and MPBs have been found in both cases [16, 17].

As previously discussed, the first step is to produce uniform films of controlled composition before aiming to CPLD. The Bi element being volatile, the stabilization of pure BFO in thin films is not straightforward, as several parasitic phases can coexist. To compensate for Bi volatility we used an enriched $Bi_{1.1}FeO_3$ target and found the range of temperature, oxygen pressure and fluence which lead to pure BFO films. The structure was studied by x-ray micro-diffraction (μXRD) and Bi/Fe ratio by Rutherford Back Scattering (RBS) versus laser fluence. A fluence of 1.72 $J{\cdot}cm^{-2}$ was identified as giving Bi/Fe = 1 with a deposition temperature of 700°C, an oxygen pressure of 0.2 mbar, a laser repetition rate of 6 Hz with a target-substrate distance of 4.5 cm [18]. RBS is an averaging technique (spot size of 2 x 2 mm^2), so to assess local uniformity of thickness and composition we turned to an Electron Probe Micro-Analyzer (EPMA) equipped for Wavelength Dispersive X-Ray Spectroscopy (WDS) (see [18] for details). A specific thin film analysis program has been used to determine BFO's composition (TFA/WDS layerf, Cameca). BFO film's density × thickness product ($\rho{\cdot}t$) and composition were simultaneously computed. Thirty measurements were realized with a beam diameter of 20 μm (20 keV, 100 nA) every 300 nm along the film. **Figure 2a** represents the ρ.t product and the weight percentages of Bi, Fe and O after self-consistent analysis of the raw data. The average value of ρ.t is 242 $\mu g/cm^2$ with a standard deviation of 5.4 $\mu g/cm^2$ equivalent to a relative variation of 2.2%. Considering the bulk BFO density (d = 8.38 g/cm^3) we find a thickness t = 289 nm ± 6.5 nm (1σ). The stability of composition is even greater along the sample. The average weight percentages and standard deviations for the Bi and Fe, transposed into atomic percentages are respectively 20.02% ± 0.08% and 19.98% ± 0.08% for an expected value of 20%. With these statistical analysis, we find that the composition dispersion is $Bi_{1.001 \pm 0.004}Fe_{0.999 \pm 0.004}O_3$ (1σ) and thickness standard deviation $\sigma(t) \leq 2.2\%$ i.e. a very good thickness and composition uniformity along the sample surface. As the aim was to measure the piezoelectric coefficient of BFO-based solid solution, we deposited BFO onto $La_{0.8}Sr_{0.2}MnO_3$ (LSMO), an epitaxial conductive oxide electrode grown on $SrTiO_3$ (001) substrate. A zoom of the θ-2θ X-ray diffraction pattern around (001) diffraction peak is shown in **Figure 2b**. Only $(00\,l)_{pc}$ (pseudo-cubic notation) oriented diffraction peaks are visible on this pattern while no parasitic phase could be detected [19]. The thickness (Pendellösung) fringes observed around both LSMO and BFO (001) reflections demonstrate the crystalline quality and the smoothness of surface and interfaces.

We choose $GaFeO_3$ (GFO) as the second member of the solid solution to be explored in order to find a MPB. GFO does not have a perovskite structure but crystallizes in a much more complex orthorhombic structure (SG Pc21n) [20]. In

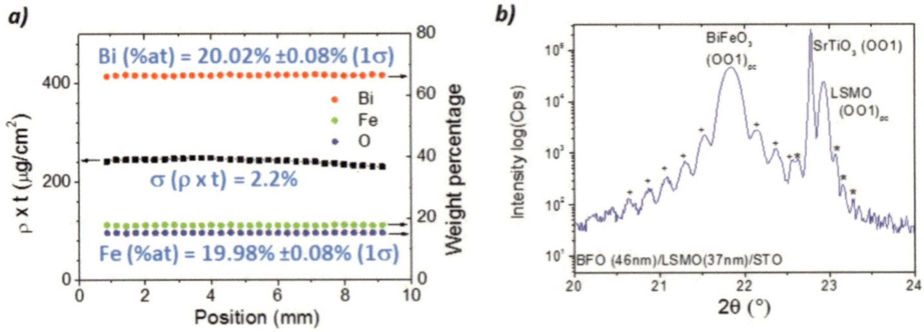

Figure 2.
(a) Evolutions of density × thickness product (ρ·t) and weight percentages of Bi, Fe and O as a function of the position (adapted from [18] with permission) and (b) X-ray diffraction pattern of BFO/LSMO/STO heterostructure.

the (1-x)BFO – (x)GFO system (BGFO) designed here, Ga^{3+} cations have incentives to substitute for Bi^{3+}. Small ions like Ga^{3+} do not occupy perovskite A-site in conventionally synthesized ceramics. However, perovskites with small A-site cations like Sc^{3+} and Mg^{2+} have recently been stabilized using high-pressure high-temperature synthesis, and Ga^{3+} is envisaged as A-site cation in this emerging field [21]. Epitaxial strain during PLD growth has longtime proven to be an alternative to high pressure synthesis for metastable phase stabilization [22]. As GFO is not a perovskite, we do not expect to obtain a solid solution at high x values, so we limited our range of investigation to $0 \leq x \leq 0.12$. In this range, the Goldsmith tolerance factor t is greater than 0.87, not too far from non-substituted BFO's tolerance factor (t = 0.89), and compatible with a distorted perovskite structure. So, it is plausible that some Ga^{3+} ions occupy the perovskite A-site in our films, although it is probable too that part of the Ga^{3+} shares the B-site together with Fe^{3+}. The Ga^{3+} substitution for Bi^{3+} being limited to 0%-12%, the BFO deposition conditions were used for GFO. WDS analysis were done along and across the composition gradient (1 point each 300 μm). Assuming the $Bi_wGa_xFe_yO_z$ formula with z = 3, cationic contents w, x, and y have been extracted from these measurements, the error being estimated to 0.005. The extracted Ga content is plotted versus position in **Figure 3a**, showing a linear increase from 0% up to 12%, in good agreement with the nominal concentrations. We note that both Bi and Fe contents decrease from 1 in pure $BiFeO_3$ to about 0.97 for 6% Ga doping. As x increases from 6–12%, Fe content goes back to about 1.0 while Bi content continue to decrease [19]. From these values, one could suspect that Ga is substituted partly for Bi and partly for Fe.

X-ray Reciprocal Space Mapping (RSM) around the $(103)_{pc}$ reflection have shown that BGFO and LSMO are epitaxial on STO (not shown here, see [19]). The lattice parameters evolutions confirm that Ga effectively enters into the BFO structure. Furthermore a characteristic splitting of the $(103)_{pc}$ reflection of BFO strained by cubic STO disappears for a Ga content 5% < x < 7%. This indicates a change of symmetry of the film and might be the signature of a MPB. Piezoelectric characterizations were made on 30 x 30 μm^2 top Pt electrodes (dc-sputter deposited via a lift-off process) using a laser scanning vibrometer (LSV model MSA-500, Polytec, V_{AC} = 1 V) (see schematic of the heterostructure **Figure 3b** top). A typical mapping of the extracted effective piezoelectric coefficient d_{33}^{eff} across one electrode, using a 3 μm laser spot size, is presented in **Figure 3b** (bottom), showing a uniform displacement. The d_{33}^{eff} coefficients were extracted from fifteen LSV measurements on each electrode and three different lines across the composition gradient were measured. **Figure 3c** shows the variation of the d_{33}^{eff} as a function of Ga content with standard deviation represented as error bars. After a slow increase of d_{33}^{eff} at low Ga content, a sharp peak centered at about 6.5% is observed.

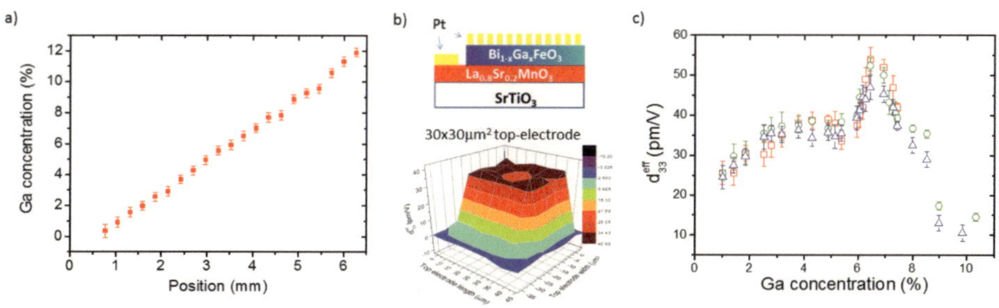

Figure 3.
(a) Ga concentration measured by WDS along the composition gradient. (b) Heterostructure schematic (top) and d_{33}^{eff} mapping of a top electrode measured by laser vibrometry (bottom). (c) BGFO d_{33}^{eff} as a function of Ga content. From [18] with permission.

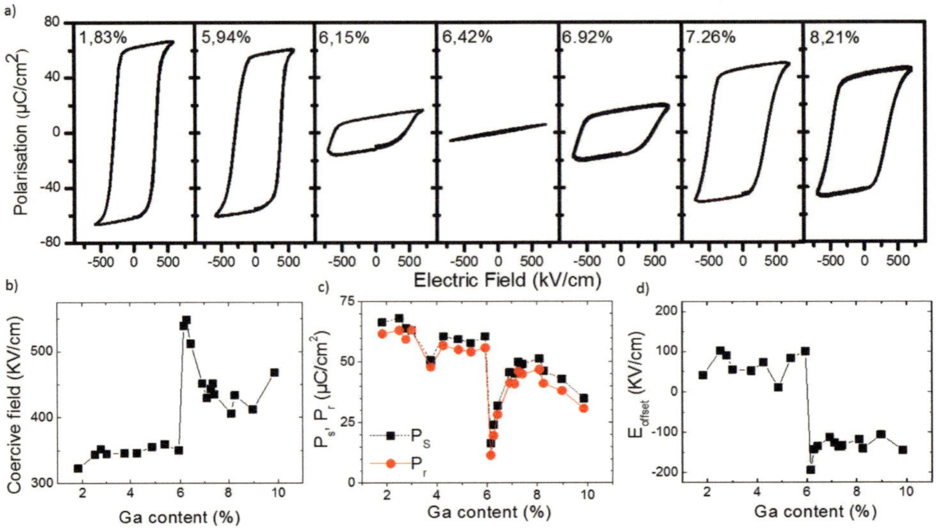

Figure 4.
(a) Ferroelectric cycles for various Ga contents measured at 77 K. the corresponding coercive fields, remanent and saturation polarizations and electric field offsets are plotted in (b), (c) and (d) respectively.

The maximum d_{33}^{eff} value, about 53 pm/V, is twice larger than the value obtained for undoped BFO. Finally, above x = 8%, d_{33}^{eff} falls to 15 pm/V. The sharp enhancement of piezoelectric properties occurring around 6.5% of Ga doping is correlated with the symmetry change observed around the same composition by RSM [19]. To confirm the presence of a MPB we looked for a change in the ferroelectric properties of BGFO with Ga content x. We used a ferroelectric tester (Radiant LC II) to measure polarization hysteresis P(E). The hysteresis cycles presented **Figure 4** were acquired at liquid nitrogen temperature.

A clear transition in the ferroelectric cycle shapes is visible **Figure 4a** as Ga content increases. As the Ga content gets over 6%, a strong increase of the coercive field (**Figure 4b**) associated to a strong decrease of both saturation and remnant polarization (**Figure 4c**) are observed, together with a change of sign of the electric field offset (hysteresis imprint **Figure 4d**). This demonstrates that a change of ferroelectric phase is occurring, correlated to the d_{33}^{eff} peak, and implies that a MPB is present in BGFO at about 6.4%. It is important to note that the d_{33}^{eff} peak is very sharp in composition and that a conventional ceramic approach with 1% doping steps would have miss it, emphasizing the power of the continuous composition spread in CPLD.

4. Interface combinatorial pulsed laser deposition

4.1 Interfaces of oxide heterostructures: the new territory

In strongly correlated complex oxides, charge, spin, orbital and lattice degrees of freedom co-exist and interplay cooperatively. In particular the complex balance between these degrees of freedom and related interactions generates a rich spectrum of competing phases in perovskites or perovskite-derived materials (e.g. high Tc superconductor, metal–insulator transitions, magnetism, ferroelectricity, piezoelectricity...). The recent progress in deposition techniques allowed the production of complex perovskites heterostructures with atomically sharp interfaces, which expanded material researcher's horizon. Fascinating phenomena and novel states of matter at complex oxide heterointerfaces have been reported. One can cite for instance the existence of high mobility 2d electron gas at $LaAlO_3$ / $SrTiO_3$ interface, even becoming superconducting at low temperature, while both materials taken separately are insulating [23, 24]. Another striking example is the transition of $CaTiO_3$ from its usual non-polar state into a high-temperature polar oxide thanks to interfacial tilt epitaxy [25]. The isomorphism of the ABO_3 perovskite oxide structure allows for a wide range of chemically modulated interfaces.

Some of the phenomenon occurring at perovskite interfaces are reported **Figure 5**. Rumpling, polar discontinuity, interfacial B-site cation environment asymmetry, BO_6 octahedral rotations are all potential levers to modulate interface properties. Their complex interplay is strongly affected by cationic substitutions and a complete and fine exploration of the possible interface compositions is required in order to identify new physics phenomena or enhanced properties. Interface CPLD (i.e. ICPLD) is a powerful tool in that respect.

One application where oxide interfaces plays a crucial role is the ferroelectric (FE) voltage tunable capacitor envisaged for future RF communication technologies (5G and Near Field Communication NFC) [27]. The relative dielectric permittivity $\varepsilon_r(E)$ of a FE has a large electric field dependence [27]. The perovskite solid solution $Ba_{1-x}Sr_xTiO_3$ (BST) is the most widely used FE in current 4G thin film parallel plates varactors because of its excellent tunability/losses compromise. New specifications for 5G and NFC (higher frequency and reduced driving voltage) call for improved varactor properties. Reducing the FE film thickness from 240 nm (4G) down to 50-100 nm range is one option toward meeting these new specifications. However, in such thin FE films the metal / FE (M/FE) interface influence is reinforced in a damaging way, due to two interfacial phenomena. The first one is the existence of FE "dead-layers" with degraded ε_r and spontaneous polarization close to the electrodes, producing an effective non-tunable interfacial capacitance [28, 29]. The second one is the increased leakage current due to insufficient Schottky Barrier Height (SBH).

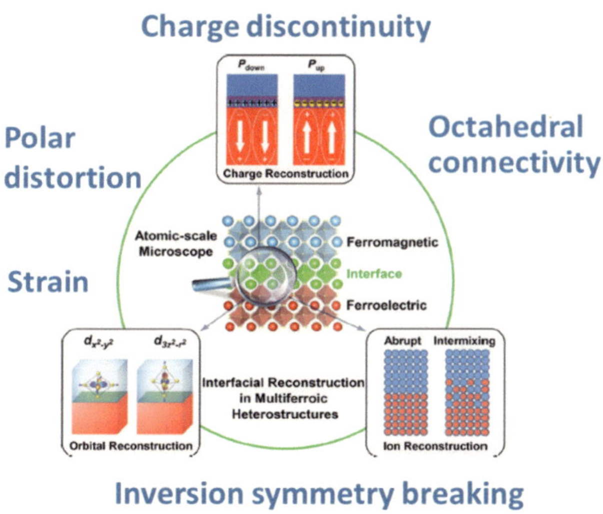

Figure 5.
Schematic of competing interactions and phenomenon at perovskite ABO_3 interfaces. From [26] with permission.

Interface engineering can be used to tailor band alignment and interface polarizability. The insertion of a thin layer with different atomic element(s) at the interface allows to manipulate the chemical bonding and promotes atomic rearrangement. Let us consider for instance the anti-displacement of anions and cations predicted at $Ba^{2+}O^{2-}$/M and $Sr^{2+}O^{2-}$/M interfaces and quantified by a rumpling parameter R [30]. R depends on the chemical bonding and is responsible for an interface dipole, which in turn modulate the SBH. Interestingly, the insertion of e.g. a single Al atomic plane at the BaO/M interface strongly affects R and SBH. Indeed, for M = Pd the SBH goes from 1.4 eV to 2.6 eV [30]. Significant rumpling has been experimentally shown for $SrTiO_3$ (STO) in contact with $La_{2/3}Sr_{1/3}MnO_3$ (LSMO), a metallic perovskite electrode, inducing a polarization in the non-ferroelectric STO [31]. The continuity of the perovskite structure through the LSMO/STO interface and its ionic character offer new ways to control electronic properties. In $La_{1-x}Sr_xMnO_3$ ($LSMO_x$), the B-site cation ratio Mn^{3+}/Mn^{4+} is determined by the A-site ratio La^{3+}/Sr^{2+}. Along [100], successive AO and BO_2 planes are polar for LSMOx and charge neutral for BST. Interfacing LSMOx with BST leads to tunable interfacial polar discontinuity which can induce lattice polar distortion and result in SBH modulation [32–34].

LSMO is a ferromagnetic (FM) half-metal, i.e. having a 100% spin-polarization at the Fermi level. For the latter reason it has been intensively studied as a spin-polarized electrode in LSMO/STO/LSMO magnetic tunnel junction (MTJ). MTJs are used e.g. as memory bits in magnetic MRAMs. The tunnel resistance depends on electrode spin-polarization and on the relative orientation of the electrode magnetic moments, with high resistance R_{AP} (resp. low resistance R_P) for antiparallel (resp. parallel) states. A 100% spin polarized electrode leads to a theoretical infinite R_{AP} which is ideal for the cited application. In LSMO/STO/LSMO, a record tunnel magneto-resistance (TMR = $(R_{AP}-R_P)/R_P$) of about 2000% was reported, but unfortunately for temperature far below the Curie temperature T_C [35]. The vast majority of the electrons tunnel from the interfaces, their spin-polarization being affected by the nature of the chemical bonding. FM correlations at manganite interfaces are known to be weaker than in bulk, causing a magnetic "dead layer" which probably explains the diminution of TMR close to T_C [36–38]. Attempts have been reported at creating a doping profile at the interfaces by inserting a 2 uc thick $LaMnO_3$ layer [39, 40] or a single uc thick $La_{0.33}Sr_{0.67}MnO_3$ [41] layer to overcome this problem with some improvement of interface magnetism but still not a full recovery of bulk properties. As for SBH and interface polarizability, multiple factors might participate to interface magnetism weakening, like charge discontinuity driven intermixing, octahedral tilt induced in the first LSMO layers by octahedral connectivity at the interface, substrate strain and so on. A combinatorial heuristic approach to the definition of interface composition is a powerful tool to help understanding all these factors interplay and to enhance the interface magnetism, SBH or interface polarization.

4.2 The LSMO/STO interfaces

Incorporating a few uc of combinatorial $LSMO_x$ ($0 \leq x \leq 1$) at the LSMO/STO interface to modulate the chemical bonding, the carrier density and the polar discontinuity could potentially induce STO lattice polar distortion, SBH modulation, as well as restoring interface ferromagnetism.

4.2.1 Ferromagnetism at STO/LSMO interface

Before producing the described ICPLD heterostructures, we first optimized the LSMO physical properties, composition and thickness uniformity. The magnetic properties of the film were then characterized versus temperature using a commercial Kerr magnetometer equipped with a cryostat (NanoMoke II, Durham Magneto Optics). As the magnetism at the LSMO/STO interface is weakened, the Curie temperature will depend on LSMO thickness for very thin films. To avoid this regime, we worked with 30 nm thick LSMO films (~80uc). Several films were deposited on TiO_2 terminated (100) STO substrates with high-pressure RHEED monitoring (Staib/TSST) at various fluence, temperature and oxygen pressure. The optimized deposition conditions leading to a T_C = 341 K were P_{O2} = 0.2 mbar, T_{sub} = 850°C, f = 5 Hz and a fluence of 0.83 J/cm^2. RHEED oscillations were clearly visible during all the deposition process implying a layer by layer growth. X-ray diffraction patterns (Θ-2Θ) showed only $(00\,l)_{pc}$ reflections with thickness fringes attesting for the crystalline quality and the surface and interface smoothness. RSM confirmed epitaxial "cube on cube" growth of LSMO on STO. The homogeneity of the films in term of composition and magnetic properties over a 1 cm^2 STO substrate was verified for thinner films, in the range where T_C is thickness dependent. A 20uc thick sample was deposited with vertical and horizontal scanning of the laser, staying in focus at the target position, and of the substrate respectively.

The Kerr magnetometer laser spot (diameter < 5 μm) was scanned on the sample surface at fixed temperatures to measure magnetic hysteresis curves in 311 points spread across the sample surface. This (x,y) scan was repeated every 2.5 K from room temperature to 350 K after thermal stabilization. Each hysteresis curve was processed in order to extract saturation and remnant magnetization (M_{sat} and M_r respectively). M(T) curves can then be reconstructed for each point on the sample surface, allowing to assess for Tc in each location. Maps reporting FM and paramagnetic (PM) areas of the sample are reproduced in **Figure 6a** (top) for various temperatures, the measurement points being indicated with black dots. The distribution of Tc is reported as a FM area percentage in **Figure 6a** bottom. Over 91% of the surface transit from FM to PM states on a temperature range less than 5 K wide (325 K < T < 330 K) and 100% inside a 10 K range. As LSMO's T_C is very thickness and composition sensitive, the tight T_C distribution indicates a good composition and thickness uniformity. We confirmed this uniformity with a WDS characterization over a 9x9 mm^2 area (25 x 25 = 625 points) of the same sample for La, Sr and Mn (JEOL 8530F). The small film thickness conjugated to the presence of Sr in the substrate did not allow to compute the composition with cationic ratios of the film. However, the WDS sensitivity is high enough to provide maps of relative variations for each element (see **Figure 6b**). The Sr map, with signal originating mostly from STO substrate, illustrates the electron beam stability (σ_{Sr} ~ 1%) which is crucial for point to point comparison. Note that the drop in the corner, corresponding to silver paste contact to evacuate the charges, was excluded from the statistical analysis. On La and Mn maps a similar slight slope is visible with a corresponding standard deviation of 4.6% and 4.9% respectively. WDS signal is strongly correlated to the thickness, therefore we can conclude that the thickness distribution is relatively tight.

The relative interface contribution to the overall magnetic signal increases as the LSMO thickness decreases. It is however difficult to predict the optimal LSMO thickness leading to an improved interface contribution detection as the overall magnetic signal also decreases with thickness. A powerful aspect of CPLD is the possibility to deposit wedge-shaped layers with continuous thickness variation using shadow-masking (see **Figure 7a**). Before inserting the ICPLD layer, we checked the thickness control on two LSMO wedges, spanning from 8 uc to 76 uc, by measuring T_C versus (x,y) and temperature. The obtained T_C are represented **Figure 7b** with standard deviation represented as error bars. Tc noticeably decreases below 30 uc with an acceleration below 20 uc. In the inset of **Figure 7b** is represented a T_C map of wedge#2, with the measured points represented as

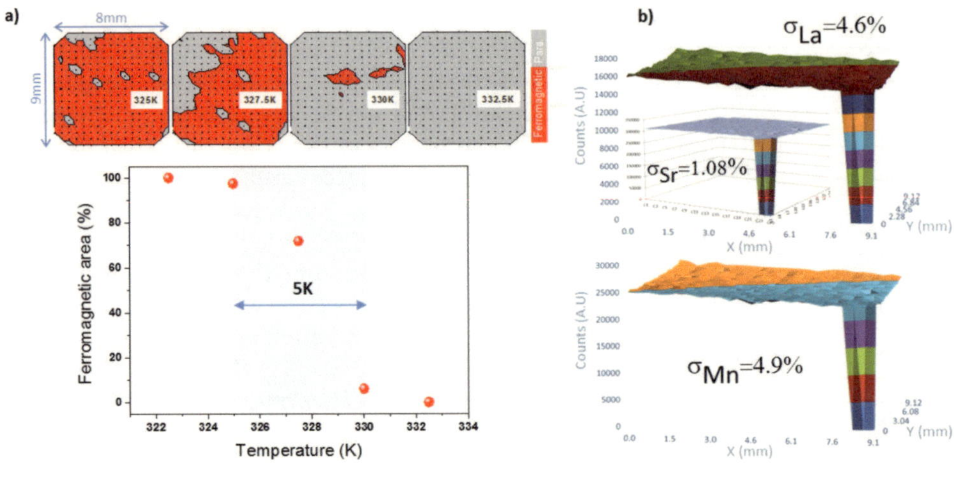

Figure 6.
(a) Magnetic state maps of LSMO 20 uc film at various temperature (top) and T_C distribution at the sample surface (bottom). (b) WDS signal for La, Sr and Mn over 9x9 mm² of the same sample.

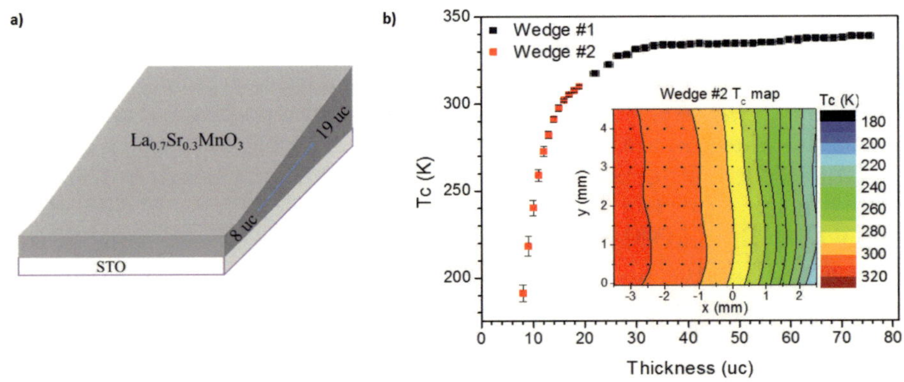

Figure 7.
(a) Schematic of LSMO wedge#2. (b) T_C versus thickness and (inset) T_C map for wedge #2.

black dots. Constant nominal thickness levels are vertical, with thickness variation along x. A 10 K color increment is used and one can see that the lines separating the adjacent areas are almost vertical, attesting the good control of thickness variation in the wedge.

To synthesize the ICPLD LSMOx layer we used $LaMnO_3$ (LMO) and $SrMnO_3$ (SMO) targets with the deposition parameters identified for LSMO, including laser and substrate stage scans. Deposition rate was evaluated using RHEED oscillations. A 3 uc thick $LSMO_x$ layer ($0 \leq x \leq 1$) was deposited onto TiO_2-terminated STO substrate, followed by a LSMO wedge with thickness variation direction perpendicular to $LSMO_x$ composition gradient. A schematic representation of this sample is represented **Figure 8a**.

M(H) cycles were acquired versus position (512 sites) and temperature (120 temperatures with 90 K < T < 340 K) automatically during a few days. Then M_{sat} and M_r were extracted for each loop, M(T) curves reconstructed and T_C estimated for each (composition x, thickness t) doublet. **Figure 8b** presents the T_C curves plotted versus t_{LSMO} for various Sr content x. One can see that the variation of Tc versus t_{LSMO} depends on x, and in particular for t_{LSMO} > 7 uc, the less Sr the more rapid is the T_C decrease. Going from LMO to SMO at t_{LSMO} = 10 uc, T_C is increased by 60 K (blue arrows **Figure 8b**). Furthermore, to reach a given Curie temperature

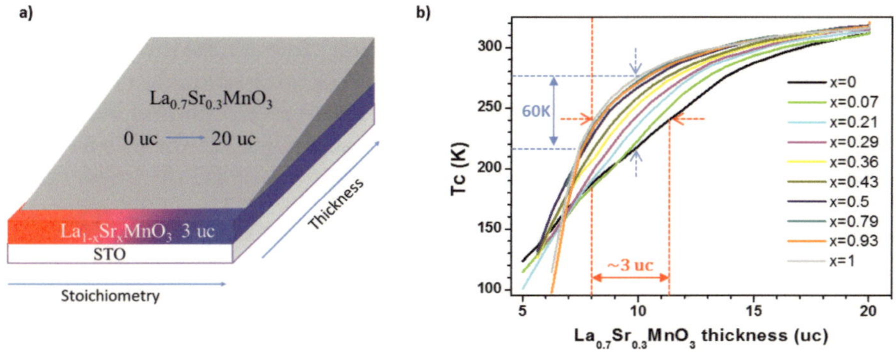

Figure 8.
(a) Schematic representation of the ICPLD $LSMO_x$ / wedge LSMO sample. (b) Curie temperature curves versus t_{LSMO} for various Sr content x in $LSMO_x$ layer.

of 240 K, one needs 8 uc of LSMO on top of SMO and more than 11 uc of LSMO on top of LMO (red arrows in **Figure 8b**). One can compare these results to the one obtained for x = 0.29 (pink curve in **Figure 8**) where the heterostructure is similar to a simple LSMO/STO interface. Inserting a 3 uc SMO layer at the LSMO/STO interface proves to be beneficial in terms of Tc for $t_{LSMO} > 7$ uc. However we observe a cross-over for $t_{LSMO} \leq 7$ uc. The Tc decrease with t_{LSMO} accelerates for Sr rich compositions, and no magnetism could be detected at $t_{LSMO} = 5$ uc for $0.29 \leq x \leq 1$. On the contrary, the lower the Sr content the higher the T_C for $0 \leq x \leq 0.21$ at $t_{LSMO} = 5$ uc. This reinforcement of FM for LMO coincides with an important increase of the coercive field to values higher than usually observed for LSMO (Hc > 300 Oe at T = 100 K). This is compatible with a second FM phase, harder than LSMO and in contact with it. LMO is antiferromagnetic (AFM) in bulk form. However, several studies reported FM LMO films on STO substrate down to 6 uc (e.g. [42]). In this article, the transition from AFM to FM has been attributed to an electronic reconstruction at the interface originating from the polar nature of the LMO. In our case the LMO layer is topped by LSMO, and it is quite possible that by proximity effect and/or stress LMO becomes FM at 3 uc thick.

4.2.2 Band alignment at LSMO/STO interface

We now turn to interface issues arising in tunable capacitors with thinned FE film i.e. the increased influence of dead ferroelectric layer on tunability and the increased leakage current. As discussed previously the insertion of a LSMOx ICPLD layer at LSMO/BST interface may increase interface polarizability and modulate SBH. In order to easily disentangle spontaneous and chemically induced polarizations we choose to work with a non-polar composition of BST i.e. STO. We deposited onto TiO_2-terminated STO substrate 38 uc of LSMO followed by 3 uc of $LSMO_x$ ($0 \leq x \leq 1$) and in the direction perpendicular to the gradient a STO wedge (3-15 uc) keeping an access to both LSMO and $LSMO_x$ with the deposition parameters described above.

A schematic of the sample structure is represented **Figure 9a**. The sample was transferred into an ultra-high vacuum atomic force microscope chamber (UHV-AFM Omicron) without breaking the vacuum. The AFM image presented in **Figure 9b** was taken about the red dot in **Figure 9a** with a total thickness of 56 uc. Terraces separated by steps of about 4 Å, i.e. one perovskite cell parameter, are clearly visible (see profile in **Figure 9c**) attesting of the layer by layer growth up to 56 uc. There exists however some 2 Å height features on the terraces indicating the probable existence of two terminations at the surface (SrO and TiO_2).

The sample was then air-exposed and inserted into a UV photoelectron spectroscopy chamber (UPS ESCALAB 250Xi Thermo Fisher) to evaluate the work function W_F as a function of position. Although the surface contamination due to air exposure prevented to extract absolute W_F values, the relative variations of W_F with Sr content and STO thickness could be determined assuming a "uniform" surface contamination. UPS spectra were taken at various x content for STO thickness ranging from 0 to 9 uc. A zoom around the emission threshold of the He II UPS spectra (He II energy 40.8 V, bias 4 V) is shown in **Figure 10a** for LSMOx (t_{STO} = 0 uc). From the thresholds one can estimated the W_F reported in **Figure 10b**. A clear continuous decrease of the work function is observed as the Sr content increases. This trend is

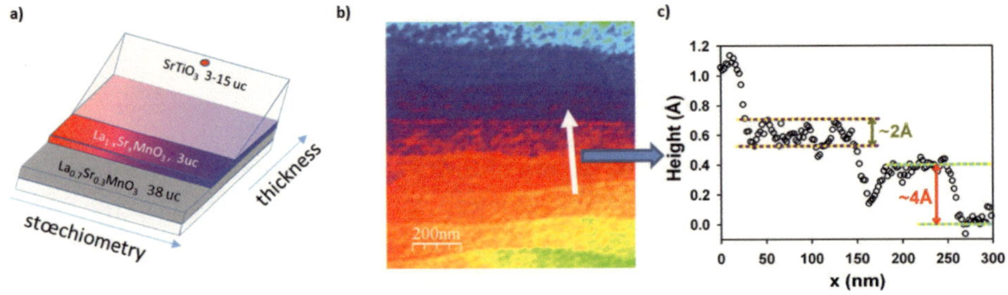

Figure 9.
(a) Schematic of the STO (sub.)/ LSMO / LSMOx / STO heterostructure. (b) UHV-AFM image. (c) Profile from the white arrow in b).

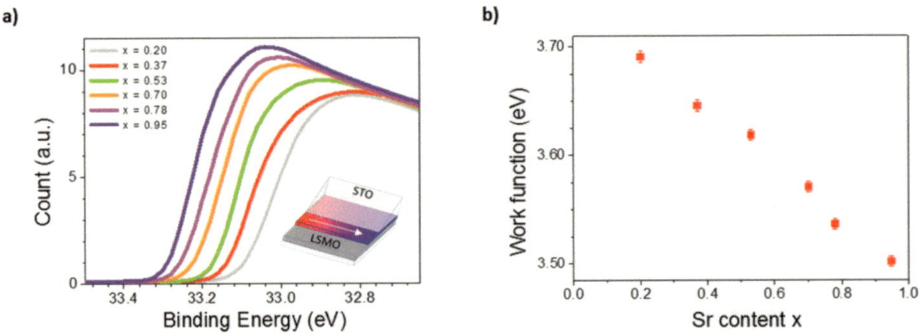

Figure 10.
(a) HeII UPS threshold for various x of air-exposed $LSMO_X$ U_{bias} = 4 V. (b) Corresponding extracted W_F versus Sr content x.

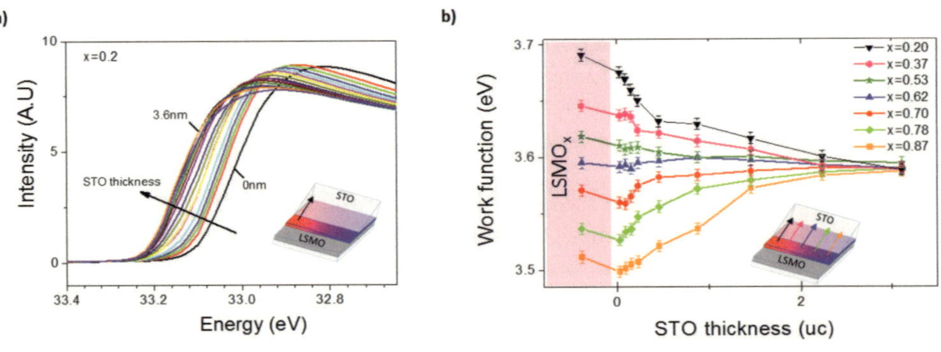

Figure 11.
(a) HeII UPS threshold for various t_{STO} at x = 0.2 U_{bias} = 4 V. (b) Corresponding extracted W_F versus t_{STO} for various Sr content x.

opposite to the downward Fermi level shift inferred from core-level XPS shift as a function of x reported in the literature [43] and seen by us (not shown).

The counter-intuitive decrease of W_F while E_F decreases too is due to the LSMOx induced charge discontinuity variation at the surface. Going from LMO to SMO, the LSMOx terminal plane changes from $Mn^{3+}O_2^{2-}$ to $Mn^{4+}O_2^{2-}$, i.e. with a surface charge per unit cell going from −1 to 0. The more negatively charged a surface is, the harder for an electron to escape from the surface, the higher the W_F [44].

The electrical nature of the contact between a metal and a semiconductor directly depends on the relative values of the metal W_F and semiconductor electronic affinity E_a for ionic semiconductor [45]. For $E_a > W_F$ an ohmic contact forms, while for $E_a < W_F$ a Schottky barrier is created. STO is generally considered an n-type ionic semiconductor with a Fermi level very close to the conduction band (i.e. $E_a \sim W_F$). As the $LSMO_x$ W_F varies the $LSMO_x$/STO contact nature might be affected. UPS spectra were acquired for various $LSMO_x$ Sr content every 200 μm along the STO wedge. A zoom of the corresponding UPS emission thresholds obtained for x = 0.2 and $0 \le t_{STO} \le 9$ uc is shown in **Figure 11a**. The threshold position varies rapidly with t_{STO} for thin STO layers then stabilizes. The W_F estimated from the UPS thresholds for various (x, t_{STO}) doublet are reported in **Figure 11b**. One can see the curves folding together toward a W_F value of about 3.58 eV (relative) corresponding to the intrinsic STO work function.

Interestingly this value is inside the range of W_F spanned within $LSMO_x$.(see pink part of **Figure 11b**). Looking at the evolution of W_F vs. t_{STO} for thicknesses up to 3 uc, there is a clear transition from a downward to an upward bending as x increases. This reflects the band bending that occurs at the $LSMO_x$ / STO interface and implies that the contact is modified from Ohmic type to Schottky type.

This result is of importance regarding the optimization of the SBH in BST FE tunable capacitor in particular, but more generally for any metal/semiconductor contacts.

5. Conclusion

In this chapter we reviewed the qualities and limitations of PLD for the synthesis of oxides in general and for its use in combinatorial PLD synthesis (CPLD) in particular. We listed some counter-actions to mitigate the PLD limitations together with the mandatory steps to take before attempting reliable CPLD synthesis, i.e. demonstrating the control of both thickness and composition over the whole sample surface. We then detailed a statistical characterization approach to reliably interpret results from CPLD libraries of compounds. An example of this approach is presented, regarding the exploration of lead-free Ga-doped $BiFeO_3$ solid solution for MPB-related piezoelectric properties enhancement. Finally we described a new interface CPLD development (ICPLD) for the exploration of functional interface libraries. This combinatorial interface synthesis approach, with continuous lateral chemical modulation of a few atomic layers, is unique to the best of our knowledge. The effectiveness of ICPLD regarding the control of interface magnetism for magnetic tunnel junctions and energy band and Schottky barrier height tuning in ferroelectric tunable capacitors was demonstrated. This shows that ICPLD is a powerful tool to accelerate heterostructures functional properties enhancement.

Acknowledgements

Authors would like to thanks J-L Longuet from CEA Le Ripault (France) for the WDS characterizations presented here and Xavier Wallart from IEMN, UMR CNRS

8520, Villeneuve d'Ascq (France), together with Pascal Andreazza from ICMN, UMR CNRS 7374, for UPS measurements.

This work was funded by Région Centre within the projects INTIM-C 2015 and FLEXIGEN.

Conflict of interest

No conflict of interest.

Author details

Jérôme Wolfman*, Beatrice Negulescu, Antoine Ruyter, Ndioba Niang and Nazir Jaber†
GREMAN Laboratory, University of Tours, Tours, France

*Address all correspondence to: wolfman@univ-tours.fr

† Current address: LCM Laboratory, CEA Leti, Grenoble, France.

References

[1] Smith HM, Turner AF. Vacuum Deposited Thin Films Using a Ruby Laser. Applied Optics.1965;4:147-148

[2] Wu MK, Ashburn JR, Torng CJ, Hor PH, Meng RL, Gao L, Huang ZJ, Wang YQ, Chu CW. Superconductivity at 93 K in a New Mixed-Phase Y-Ba-Cu-O Compound System at Ambient Pressure. Physical Review Letters. 1987;58:908-910. DOI: 10.1103/PhysRevLett.58.908

[3] Chu CW. Cuprates—Superconductors with a Tc up to 164 K. In: Rogalla H, Kes PH, editors. 100 years of superconductivity. Boca Raton: CRC Press/Taylor & Francis Group. pp. 244-253. ISBN 9781439849484.

[4] Dijkkamp D, Venkatesan T, Wu XD, Shaheen SA, Jisrawi N,Min-Lee YH, McLean WL and Croft M. Preparation of Y-Ba-Cu oxide superconductor thin films using pulsed laser evaporation from high Tc bulk material. Appl. Phys. Lett. 1987;51:619. DOI: 10.1063/1.98366

[5] Chrisey DB, Hubler GK. Pulsed laser deposition of thin films. New York: Wiley; 1984. ISBN 0-471-59218-8

[6] Ohnishi T., Shibuya K., Yamamoto T. and Lippmaa M. Defects and transport in complex oxide thin films. J. Appl. Phys. 2008;103:103703. DOI: 10.1063/1.2921972

[7] Guozhen L, Wolfman J, Autret-Lambert C, Sakai J, Roger S, Gervais M, Gervais F. Microstructural and dielectric properties of $Ba_{0.6}Sr_{0.4}Ti_{1-x}Zr_xO_3$ based combinatorial thin film capacitors library. Journ. Appl. Phys. 2010;108:114108

[8] Daumont CJM, Simon Q et al. Tunability, dielectric, and piezoelectric properties of $Ba_{1-x}Ca_xTi_{1-y}Zr_yO_3$ ferroelectric thin films. Journ. Appl. Phys. 2016;119:094107. DOI: 10.1063/1.4942924

[9] Trolier-McKinstry S, Muralt P, J. Thin Film Piezoelectrics for MEMS. J. Electroceram. 2004;12:7-17

[10] Jaffe B, Cook WR, Jaffe H. Piezoelectric Ceramics. London : Academic Press, 1971. p136. ISBN-0123795508

[11] Davis M. Picturing the elephant: Giant piezoelectric activity and the monoclinic phases of relaxor-ferroelectric single crystals. J. Electroceram. 2007;19:25

[12] Ahart M, Somayazulu M, Cohen RE, Ganesh P, Dera P, Mao H-K, Hemley RJ, Ren Y, Liermann P, Wu Z. Origin of morphotropic phase boundaries in ferroelectrics. Nature 2008;451:545-548 DOI : 10.1038/nature06459

[13] Noheda B, Cox DE, Shirane G, Gonzalo JA, Cross LE, Park S-E. A monoclinic ferroelectric phase in the $Pb(Zr_{1-x}Ti_x)O_3$ solid solution. Appl. Phys. Lett. 1999;74: 2059. DOI: 10.1063/1.123756

[14] Kaczmarek W and Pajak Z, Differential thermal analysis of phase transitions in $(Bi_{1-x}La_x)FeO_3$ solid solution. Solid State Commun. 1975;17: 807-810. DOI: 10.1016/0038-1098(75)90726-7

[15] Roginskaya YE, Tomoshpolskii YY, Venevstev YN, Petrov VM, Zhdanov GS. The Nature of the Dielectric and Magnetic Properties of $BiFeO_3$. Sov. Phys.-JETP 1966;23:47

[16] Wei Y, Wang X, Zhu J, Wang X, Jia J. Dielectric, ferroelectric, and piezoelectric properties of $BiFeO_3$–$BaTiO_3$ ceramics. J. Am. Ceram. Soc. 2013;96: 3163-3168. DOI: 10.1111/jace.12475

[17] Woodward DI, Reaney IM, Eitel RE, Randall CA. Crystal and domain

structure of the $BiFeO_3$–$PbTiO_3$ solid solution. J. Appl. Phys. 2003 ;94 :3313. DOI : 10.1063/1.1595726

[18] Jaber N, Wolfman J, Daumont C, Négulescu B, Ruyter A et al. Laser fluence and spot size effect on compositional and structural properties of $BiFeO_3$ thin films grown by Pulsed Laser Deposition. Thin Solid Films 2017;634:107-111. DOI: 10.1016/j.tsf.2017.05.003

[19] Jaber N, Wolfman J, Daumont C, Négulescu B et al. Enhancement of piezoelectric response in Ga doped $BiFeO_3$ epitaxial thin films. J. Appl. Phys. 2015;117:244107; doi: 10.1063/1.4923217

[20] Abrahams SC, Reddy JM, Bernstein JL. Crystal structure of piezoelectric ferromagnetic gallium iron oxide. J. Chem. Phys. 1965;42:3957. DOI: 10.1063/1.1695868

[21] Belik A, Yi W. High-pressure synthesis, crystal chemistry and physics of perovskites with small cations at the A site. J. Phys.: Condens. Matter 2014;26:163201. DOI: 10.1088/0953-8984/26/16/163201

[22] Norton DP, Chakoumakos BC, Budai J, Lowndes DH, Sales BC, Thompson JR, Christen D. Superconductivity in $SrCuO_2$-$BaCuO_2$ superlattices: formation of artificially layered superconducting materials. Science 1994;265:2074-2077. DOI: 10.1126/science.265.5181.2074

[23] Ohtomo A, Hwang HY, A high-mobility electron gas at the $LaAlO_3$/$SrTiO_3$ heterointerface. Nature 2004;427;423-426. DOI: 10.1038/nature02308

[24] Reyren R, Thiel S et al. Superconducting interfaces between insulating oxides. Science 2007;317:1196-1199. DOI: 10.1126/science.1146006

[25] Yuan Y et al. Three-dimensional atomic scale electron density reconstruction of octahedral tilt epitaxy in functional perovskites. Nature Com. 2018;9:5220. DOI: 10.1038/s41467-018-07665-1

[26] Huang W, Yin Y, Xiaoguang L. Atomic-scale mapping of interface reconstructions in multiferroic heterostructures. Appl. Phys. Rev. 2018;5 :041110. DOI : 10.1063/1.5053426

[27] Tiggelman M et al. On the trade-off between quality factor and tuning ratio in tunable high-frequency capacitors. IEEE trans. Elec. Dev. 2009;56:2128-2136. Doi: 10.1109/TED.2009.2026391.

[28] York R, Tunable Dielectric Materials in : M. Steer and W.D. Palmer editors, Multifunctional Adaptive Microwave Circuits and Systems, New York: Scitech: 2009, chap. 4

[29] Lichtensteiger C et al. Ferroelectricity in ultrathin capacitors in: Pacchioni G, Valeri S editors, Oxide Ultrathin Films: Science and Technology, Weinheim: WYLEY;2012 Chap.12 ISBN 978-3-527-33016-4

[30] Nuñez M and Nardelli MB. First-principles theory of metal-alkaline earth oxide interfaces. Phys. Rev. B 2006;73:235422. DOI: 10.1103/PhysRevB.73.235422

[31] Wang Z et al. Anomalously deep polarization in (001) interfaced with an epitaxial ultrathin manganite film. Phys. Rev. B 2016;94;155307. DOI: 10.1103/PhysRevB.94.155307

[32] Yu P, Chu Y-H, Ramesh R. Oxide interfaces: pathways to novel phenomena. Mat. Today 2012;15:320-327. ISSN:1369 7021

[33] Koohfar S et al. Structural distortions at polar manganite interfaces. Phys. Rev. B 2017;96:024108. DOI: 10.1103/PhysRevB.96.024108

[34] Hikita Y et al. Termination control of the interface dipole in $La_{0.7}Sr_{0.3}MnO_3$/ Nb:$SrTiO_3$ (001) Schottky junctions. Phys. Rev. B 2009;79:073101 (2009) DOI: 10.1103/PhysRevB.79.073101

[35] Werner R, Petrov A Yu, Alvarez Miño L, Kleiner R, Koelle D, Davidson B A. Improved tunneling magnetoresistance at low temperature in manganite junctions grown by molecular beam epitaxy. Appl. Phys. Lett. 2011 ;98 ;162505. DOI : 10.1063/1.3581885

[36] Sun J Z, Abraham D W, Rao R A, Eom C B. Thickness-dependent magnetotransport in ultrathin manganite films. Appl. Phys. Lett. 1999;74:3017. DOI: 10.1063/1.124050

[37] Bibes M, Balcells L, Valencia S, Fontcuberta J, Wojcik M, Jedryka E, Nadolski S. Nanoscale Multiphase Separation at $La_{2/3}Ca_{1/3}MnO_3$/ $SrTiO_3$ Interfaces. Phys. Rev. Lett. 2001;87;067210. DOI: 10.1103/ PhysRevLett.87.067210

[38] Tebano A, et al. Evidence of orbital reconstruction at interfaces in ultrathin $La_{0.67}Sr_{0.33}MnO_3$ films Phys. Rev. Lett. **100**, 137401 (2008).

[39] Yamada H, Ogawa Y, Ishii Y, Sato H, Kawasaki M, Akoh H, Tokura Y. Engineered interface of magnetic oxides. Science 2004;305:64-648. DOI: 10.1126/ science.1098867

[40] Kavich JJ et al. Nanoscale suppression of magnetization at atomically assembled manganite interfaces: XMCD and XRMS measurements. Phys. Rev. B 2007;76:014410. DOI: 10.1103/ PhysRevB.76.014410

[41] Boschker H et al. Preventing the reconstruction of the polar discontinuity at oxide heterointerfaces. Adv. Funct. Mater. 2012;22:2235-2240. DOI: 10.1002/adfm.201102763

[42] Wang R et al. Imaging and control of ferromagnetism in $LaMnO_3/SrTiO_3$ heterostructures. Science 2015 ;349 :716-719. DOI: 10.1126/science.aaa5198

[43] Saitoh T, Bocquet AE et al. Electronic structure of $La_{1-x}Sr_xMno3$ studied by photoemission and x-ray-absorption spectroscopy. Phys. Rev. B 1995;51:13942. DOI: 10.1103/ PhysRevB.51.13942

[44] Ryan Jacobs , John Booske , and Dane Morgan. Understanding and Controlling the Work Function of Perovskite Oxides Using Density Functional Theory. *Adv. Funct. Mater.* **2016**, *26*, 5471-5482

[45] R. Tung, Appl. Phys. Rev. 1, 011304 (2014) https://doi. org/10.1063/1.4858400

Chapter 2

Pulsed Laser Deposition of Transition Metal Dichalcogenides-Based Heterostructures for Efficient Photodetection

Deependra Kumar Singh, Karuna Kar Nanda and Saluru Baba Krupanidhi

Abstract

From the past few decades, photodetectors (PDs) are being regarded as crucial components of many photonic devices which are being used in various important applications. However, the PDs based on the traditional bulk semiconductors still face a lot of challenges as far as the device performance is concerned. To overcome these limitations, a novel class of two-dimensional materials known as transition metal dichalcogenides (TMDCs) has shown great promise. The TMDCs-based PDs have been reported to exhibit competitive figures of merit to the state-of-the-art PDs, however, their production is still limited to laboratory scale due to limitations in the conventional fabrication methods. Compared to these traditional synthesis approaches, the technique of pulsed laser deposition (PLD) offers several merits. PLD is a physical vapor deposition approach, which is performed in an ultrahigh-vacuum environment. Therefore, the products are expected to be clean and free from contaminants. Most importantly, PLD enables actualization of large-area thin films, which can have a significant potential in the modern semiconductor industry. In the current chapter, the growth of TMDCs by PLD for applications in photodetection has been discussed, with a detailed analysis on the recent advancements in this area. The chapter will be concluded by providing an outlook and perspective on the strategies to overcome the shortcomings associated with the current devices.

Keywords: two-dimensional semiconductors, transition metal dichalcogenides, pulsed laser deposition, photodetectors

1. Introduction

Photodetectors (PDs) are the optoelectronic devices which convert incident optical signals into electrical outputs through the phenomenon of light-matter interaction, which can be processed by the conventional read-out electronics. PDs form the basis of many vital components present in numerous electronic and optoelectronic devices as they find applications in a broad range of fields such as in photovoltaics [1, 2], military and defense technology [3], optical communication [4], remote sensing [3], biomedical imaging [5], environmental and ozone layer monitoring [4], and so on. Therefore, highly efficient photodetection has become very crucial for the industrial and scientific communities. With advancements in the matured technology of three-dimensional (3D) semiconductors [6–17] such as gallium nitride (GaN), zinc oxide (ZnO), indium gallium nitride (InGaN), indium nitride (InN), gallium oxide (Ga_2O_3), gallium arsenide (GaAs), silicon (Si), aluminum gallium nitride (AlGaN), germanium (Ge), mercury cadmium telluride (HgCdTe), gallium antimonide (GaSb), and so forth, high-performance PDs sensitive to wavelengths in the entire ultraviolet (UV)-far infrared (FIR) have been successfully fabricated. However, further advances in these PDs are hindered by the certain drawbacks encountered due to the intrinsic limitations in 3D semiconductors such as lower charge carrier mobilities, low light absorption properties, presence of dangling bonds at the interface, high fabrication costs involved, and so forth [18]. Thus, it is crucial to explore alternate materials, which can overcome the above-mentioned drawbacks for the development of multifunctional PDs.

The successful delamination of graphene in the revolutionary work by Geim and Novoselov in 2004 [19] ignited a plethora of research, in the field of two-dimensional (2D) layered materials and their heterostructures [20–22]. In the recent years, the amount of research focused on these layered materials has increased multifold. A layered material is nothing but an ultrathin phase of a material, scaled down to the level of atomic thickness, and is characterized by weak inter-layer van der Waals (vdW) forces and a strong intra-layer covalent interaction [23]. This makes these ultrathin materials possess electronic and optoelectronic properties such as band gap, mobility, etc., that are thickness-dependent [24], and thus, their novel chemical and physical characteristics pave a way towards the unexplored areas, both in the fields of fundamental research as well as engineering applications. In spite of possessing unparalleled electronic and optoelectronic properties such as high carrier mobility, dangling bonds-free surface, large current carrying capacities, excellent mechanical properties, the zero band gap or the gapless electronic structure of graphene [25] limits its use in realization of practical applications-based PDs, which demand switching behavior or in other words, a definite on/off state. This has led to the exploration of graphene alternatives with a substantial band gap, and researchers and scientists across the world have resurrected a class of conventional 2D materials known as the transition metal dichalcogenides (TMDCs), characterized by low-fabrication cost, chemical stability, earth-abundance and environment-friendly properties. Some of the well-studied TMDCs are molybdenum disulfide (MoS_2), tungsten disulfide (WS_2), molybdenum diselenide ($MoSe_2$), molybdenum ditelluride ($MoTe_2$), tungsten diselenide (WSe_2), and so on [26–28]. As of now, semiconductors of the TMDC family have enabled tremendous accomplishments in the field of photodetection, such as from monofunctional to multifunctional PDs, from homogeneous to hybrid 2D semiconductors-based PDs, and from rigid to flexible electronic devices [29–31].

Regardless of the efforts of the researchers and scientists, some common challenges are still being faced related to fabrication and the performance of these TMDCs-based devices [18]. The challenges include growth of high-quality crystals, controlling the morphology and the thickness, scaling up the growth for industrial scale production, optimizing the device architectures, and so forth. To address these challenges, pulsed laser deposition (PLD) has emerged as a perfect tool for the synthesis of TMDCs. With the use of PLD, the actualization of high quality and wafer scale synthesis of TMDCs has become possible [32]. Eminently, PDs based on the PLD-synthesized TMDCs have exhibited competitive device performance parameters when compared with the commercial PDs, and thus, offer great opportunities towards the next generation photonics. In the subsequent section, we will give an introduction about TMDCs and their properties. Afterwards, the fundamentals of PLD will be discussed in detail, followed by the recent advancements in the PLD-grown TMDCs for photodetection application. Finally, we will conclude by highlighting the unresolved problems and suggest future perspectives in this evolving field of optoelectronics.

2. Transition metal dichalcogenides (TMDCs)

TMDCs are denoted by the general formula of MX_2 where M and X represent a transition metal (Mo, Nb, W, Hf, Ti, and so on), and a chalcogen (S, Te, and Se), respectively. In the periodic table, groups IV-X belong to the transition metals and they contain different number of d-electrons. Thus, the difference in the valance d-electrons of different transition metals gives rise to the different electronic properties such as metallic, semiconducting, and superconducting [33]. TMDCs exist in a layered form at the atomic level, containing one or few monolayers. **Figure 1(a)** shows a schematic depicting the layered structure of MoS_2.

A TMDC can exist in various structural phases which are a result of the different co-ordinations of the transition metal atom. The most common structural phases in which the TMDCs crystallize are the trigonal prismatic (2H) and the octahedral (1T) phase. These crystal phases can also be seen in the terms of different stacking sequences of the atoms (as a representative result, **Figure 1(b)** shows crystal structure of MoS_2). The three atomic planes i.e. chalcogen– metal–chalcogen form the individual layers of TMDCs. The 2H phase corresponds to an ABA stacking, whereas, the 1T phase is characterized by an ABC stacking order. For most of the bulk TMDCs (MoS_2, $MoSe_2$, $MoTe_2$, WS_2, WSe_2, etc.), the 2H phase is thermodynamically more stable than the metastable 1T phase. Tungsten ditelluride (WTe_2) shows an exception where the most stable phase is the orthorhombic ($1T_d$ phase) at room temperature [33].

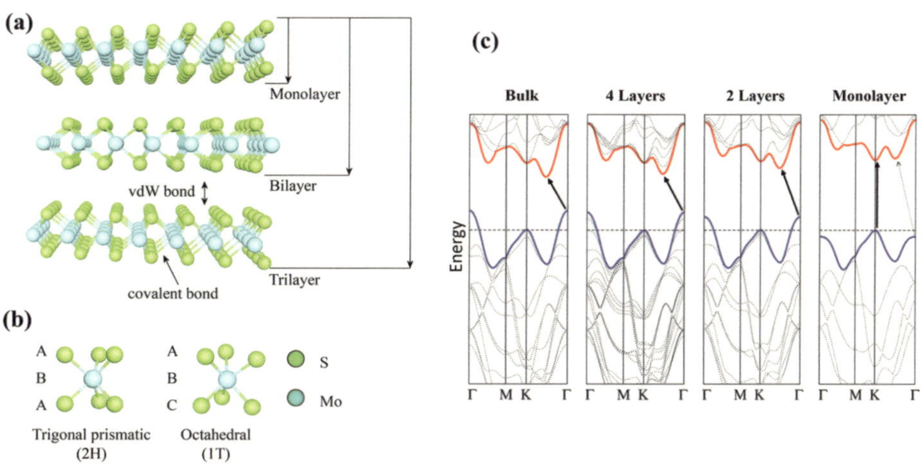

Figure 1.
(a) Schematic showing a monolayer of a TMDC, where the atoms of transition metal are bonded through covalent bonds with the chalcogen atoms. These individual monolayers are stacked and held together by vdW forces to form the bulk structures. (b) Crystal structures of layered MoS_2 with different stacking sequences as shown to form the two most common phases: trigonal prismatic (2H) and octahedral (1T). Figure is adapted from Ref. [23]. (c) Transformation of the band structure of 2H phase of MoS_2 calculated by first principles from bulk to a single layer. Figure is adapted and reproduced with permission from Ref. [34].

The assortment of chemical compositions as well as the different crystal structures of TMDCs results in varying band structure characters, which in turn lead to a wide range of electronic properties. In the thermodynamically most stable 2H phase, MoS_2, WS_2, $MoSe_2$, and WSe_2 show semiconducting behavior [33]. These semiconducting properties accentuated these TMDCs as potential 2D materials for next generation electronic devices. As a representative result, the basic characteristics of the band structure of MoS_2 has been shown in **Figure 1(c)**. The transformation of the band structure as calculated from density functional theory (DFT) for 2H-MoS_2 upon increasing its thickness from monolayer form to the bulk has been shown in **Figure 1(c)**. The positions of the conduction and valence band edges change with increasing the number of layers of MoS_2, and the direct band gap in the monolayer form changes into an indirect band gap in the bulk material [33]. The calculated value of the band gap of the monolayer 2H-MoS_2 is ~1.89 eV [35]. The experimentally observed value for the electronic band gap of 2H-MoS_2 in its monolayer form is 2.15 eV [36]. Notably, the conduction band minimum and the valence band maximum are situated at the two inequivalent high-symmetry points, which represent the corners of the hexagonal Brillouin zone [33]. This attribute is common between monolayer 2H phase MoS_2 (and the other group VI single-layer 2H phase TMDCs) and graphene and allows the observation of potential valleytronics applications.

Through the persistent efforts over the last decade, researchers have developed several techniques to produce ultrathin TMDCs. In principle, these techniques can be broadly classified into two categories. The first one is to produce TMDCs by thinning bulk crystals, which is called as a top-down technique. Different types of exfoliations (mechanical, chemical, etc.) fall under this category. Since different layers of 2D materials are held together by vdW interactions, therefore, the interlayer bonding is weak. Thus, under any external perturbations, bulk 2D materials are readily processed into their few-layered forms. The second category is to produce TMDCs through a bottom-up approach, where constituent atoms and molecules are assembled together to form continuous layers. These mainly include chemical vapor deposition (CVD), atomic layer deposition (ALD), magnetron sputtering, molecular beam epitaxy (MBE) and PLD.

As mentioned above, researchers have developed several methods to fabricate TMDCs. In spite of the substantial progress, none of the above techniques can meet the comprehensive demands for industrial-scale production, as in the terms of process simplicity, good scalability, excellent homogeneity and continuity, high quality of the products, high compositional and thickness control, low cost for mass production and higher safety. The techniques such as exfoliation, CVD and ALD suffer from a huge drawback that the growth process occurs in absence of high vacuum, which generally leads to unclean interfaces, and therefore, one has to compromise with the device performance. Moreover, exfoliation and CVD are further associated with low product yield as the films formed are non-continuous and in the range of several microns. Sputtering and MBE are much more sophisticated in terms of the interface quality due to the involvement of ultrahigh vacuum, however, sputtering is characterized by poor surface quality of the films whereas MBE poses drawbacks such as bulky and expensive setups, time-consuming growths, and limitations in terms of substrates. Such limitations further hinder the utilization of these growth methods for industrial scale fabrication of devices. Thus, research and development towards a potentially competent approach which can address the above issues is greatly required.

PLD is a synthesis technique where a high-power pulsed laser beam is focused on a target, which results in the vaporization of the material in the form of a plasma plume and this material gets deposited on a substrate in the form of a thin film.

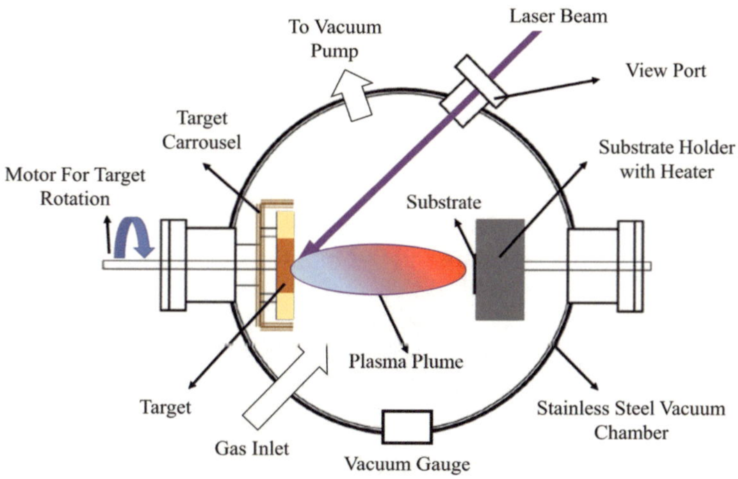

Figure 2.
Schematic view of a typical PLD process. Adapted from Ref. [37].

A typical illustration of a PLD process is shown in **Figure 2**. Compared with the conventional methods discussed above, PLD exhibits the following advantages:

- PLD is the most versatile growth method where a focused and high-energy pulsed laser ionizes almost all types of materials, because of the generation of instantaneous and localized high temperatures up to tens of thousands of ^{O}C on the target's surface. Therefore, majority of the materials can be ablated to form a plasma, which carries out the deposition. In addition, PLD exhibits an excellent compatibility with different substrates which provides numerous routes for the construction of heterojunctions-based devices.

- PLD is highly scalable, because the plasma plume can be readily positioned and directed just by adjusting the external optical path of the laser beam. Consequently, PLD is very much suitable for the growth of wafer-scale and uniform TMDCs for practical industrial production. Serna *et al.* [38] have successfully fabricated continuous bilayers of MoS_2 on sapphire with a diameter up to ~50.8 mm using PLD. Singh *et al.* [39] have also synthesized centimeter-scale MoS_2 thin films on various substrates.

- The substrate temperature required for the PLD growth of TMDCs is relatively low when compared with techniques like CVD and sputtering because the species (atoms, molecules, ions, etc.) ablated by the focused pulsed laser possess a very high energy, and therefore, can freely migrate on the substrate's surface. Therefore, direct growth of TMDCs on substrates that are intolerable towards high temperature can be achieved with PLD. For instance, Singh *et al.* [39] have successfully fabricated MoS_2 on InN at a low substrate temperature of 450°C. It may be noted that low temperature deposition is required for InN as it dissociates above 500°C.

- PLD is a clean, highly efficient, safe, and highly controlled deposition technique where the product is highly continuous and uniform. Due to the ultrahigh vacuum growth conditions, the products of PLD are clean and contamination-free. Furthermore, Siegel *et al.* [32] fabricated MoS_2 on centimeter-scale sapphire substrates using PLD, varying the thickness from 60 monolayers down to a single monolayer, just by tuning the number of laser pulses.

In the past few years, TMDCs have elicited tremendous research interest, owing to their novel properties in the 2D form, which has triggered a spark in the growth of TMDCs using PLD [18]. In this part of the chapter, a few reports describing the chronological developments in the area of PLD grown TMDCs have been briefly discussed. One of the earliest works on PLD deposited MoS_2 was reported by Zabinski *et al.* [40], where they have prepared PbO-MoS_2 thin films for tribological applications. Other early reports on MoS_2 thin films by PLD include growth of amorphous MoS_2 by McDevitt *et al.* [41] and MoS_2 coatings for friction-related studies by Mosleh *et al.* [42]. The major advancements in this area have been achieved in the past few years. In 2010, Fominski *et al.* [43] have experimentally studied the fabrication of $MoSe_x$ thin films with varying compositions obtained by PLD in vacuum condition and in presence of different rarefied gases. They also developed a process-based mathematical model which played a dominant role on the chemical composition of these thin films. Loh *et al.* [44] in 2014 fabricated MoS_2 on different metals such as Ag, Ni, Al, and Cu by PLD. In 2015, a significant step towards the fabrication of transfer-free TMDC-based PDs was demonstrated by Serrao *et al.* [45] in which MoS_2 was directly deposited on substrates like sapphire, SiC-6H and GaN. The first significant work towards the wafer-scale growth of TMDCs by PLD was done by Siegel *et al.* [32] who reported growth of centimeter-scale MoS_2 thin films of varying thickness (from monolayer to 60 monolayers). Growth of other members of the TMDC family such as WS_2 was also investigated simultaneously. In 2015, Loh *et al.* [46] synthesized WS_2 thin films on Ag substrates by PLD and reported the thickness dependent Raman and PL spectra. However, the obtained WS_2 was having a mixed phase (1T and 2H). In subsequent works, Yao *et al.* [47] obtained the pure 2H phase WS_2 directly by PLD on insulating SiO_2/Si substrates. The growth of selenides via PLD usually suffers from a large number of Se vacancies. Therefore, a two-step growth method was adopted for production of $MoSe_2$ [48]. This included deposition of MoO_3 film via PLD, followed by its selenization. Later on, Mohammed *et al.* [49] achieved 1–8 monolayers of WSe_2 via single-step deposition. This was achieved through a hybrid PLD cluster, where a tungsten target was ablated by the laser beam and selenium vapors were synchronously provided from an effusion cell by thermal evaporation. In 2018, a single-step PLD approach was used by Seo *et al.* [50] to deposit WSe_2 on Al_2O_3 and SiO_2/Si substrates by using a Se-rich target. Lately, Gao *et al.* [51] demonstrated a two-step synthesis route to fabricate 2D WTe_2, which included PLD of amorphous WTe_2 target followed by annealing treatment of the thin films in a Te atmosphere. These reports suggest that the technique of PLD can be suitably applied for the successful production of various TMDCs. In the next section, we will discuss in detail about the fundamentals associated with PLD.

3. Basics of pulsed laser deposition

PLD falls under the category of physical vapor deposition and is a method used to synthesize materials (generally thin films) in an ultrahigh vacuum environment. The development in the field of laser-assisted film growth can be traced back to 1960, after the successful technical realization of the first laser by Maiman [52]. Following on from there, from just being a growth method for fundamental laboratory research, PLD has moved on to become a technique employed in industries. A typical PLD system mainly consists of a laser (usually an excimer laser), an optical path system consisting of apertures, lenses and reflectors, and a stainless-steel growth chamber equipped with gas paths, vacuum pumps, vacuum gauges, and a heating source [37]. The basic principle behind a PLD process is that a high-intensity pulsed laser interacts with the target or the source material (called ablation) and produces a plasma plume of the target material [37]. The formation of plasma involves a sequence of complex phenomena such as collision, localized heating and subsequent ionization of atoms and molecules. Afterwards, the plasma plume expands, travels downstream, condenses on the substrates and finally crystallizes into the desired materials [37]. **Figure 3** shows the PLD setup located in Materials Research Centre, Indian Institute of Science, Bangalore, India.

The major advantage of a PLD system is that the laser can be operated from outside the vacuum chamber. Thus, just by changing the optical paths of the laser beam, a single laser source can be used for multiple deposition systems. All the other components such as the target carousel, substrate holder, heater, vacuum gauges, and so on are mounted in the vacuum chamber. A set of optical components such as apertures and mirrors are used to focus the pulsed laser beam over the target surface. Therefore, a variety of materials (semiconductors, metals and insulators) can be grown by PLD just by optimizing the growth parameters and by incorporation of different gases during the process in a controlled manner. With this, the exact stoichiometry of the target material can be copied down onto the substrates, which is one of the major benefits of PLD over other deposition techniques. Some of the important parameters associated with PLD have been described below.

Laser source: A krypton fluoride (KrF) laser is a type of excimer laser, and with a wavelength of 248 nm, it is a deep UV laser which is commonly used for the growth of various thin films as the absorption spectrum of most of the inorganic materials lies in the range of 200–400 nm. Typically, excimer lasers contain a mixture of two gases: a noble gas such as argon, xenon or krypton; and a halogen such as chlorine or fluorine. Under suitable conditions of stimulation and pressure, an excimer molecule is created, which decays via a stimulated emission and a coherent beam of stimulated radiation is emitted in the UV range. The continuous emission is

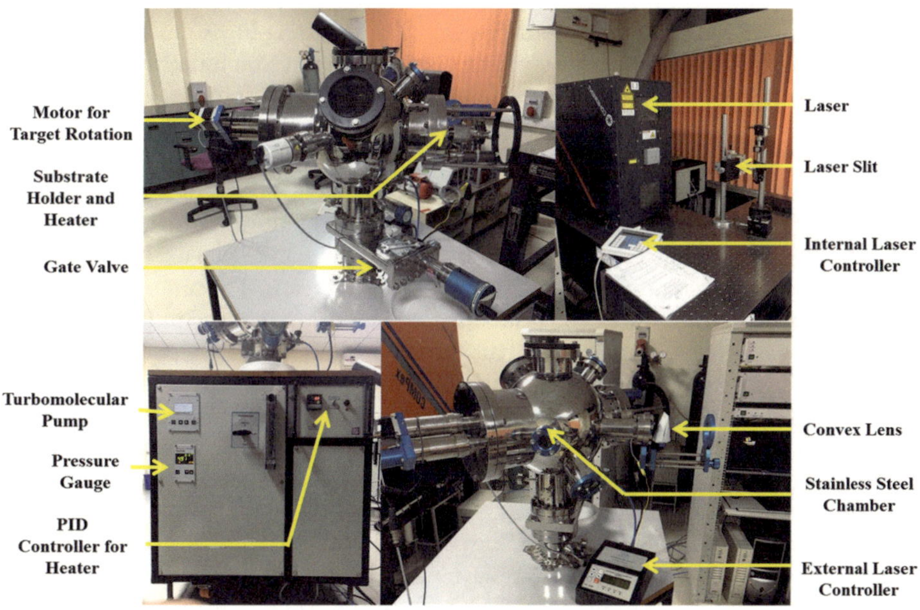

Figure 3.
PLD setup in Materials Research Centre, Indian Institute of Science, Bangalore, India.

then converted into a pulse by various discharge mechanisms and a pulse width of ~10–20 nanoseconds (ns) can be achieved.

Laser fluence: The laser fluence or laser energy density is defined as the laser output energy per unit area and is a very important parameter which decides the proper ablation of the target where the laser beam interacts with the target. A minimum threshold laser fluence is required to carry out the proper ablation process, otherwise, only evaporation takes place. The plume formation depends upon the target conditions such as its density, porosity, morphology, and compositional impurities as well as the laser conditions such as laser pulse duration and laser pulse width. If the laser fluence is much above the threshold value, crystallographic defects and damage can occur in the deposited thin film because of the bombardment by the ablation particles possessing high kinetic energy. Also, it can lead to macroscopic particles ejection during the process of ablation, particulate formation on thin films as well as back-sputtering of species from the deposited thin film. Various mechanisms have been proposed for the formation of particulates and several methods have been devised to minimize these effects [53, 54].

Laser-target interactions: The three main processes taking place during the laser-target interaction are: (i) the laser beam interacts with the surface of the target and gets absorbed into surface layer; (ii) the removal of atomic species from the material is done by vaporization of the surface region in a non-equilibrium state; (iii) afterwards, rapid vaporization further produces a recoil pressure, which leads to the expulsion of the molten pool and produces the plasma plume, and the formed plasma is a collection of electrons, neutral atoms, ions, etc. Therefore, the absorption process is highly dependent on the target properties as well as the laser characteristics. Also, this absorption process is different for metals, insulators and semiconductors [55, 56]. When the laser beam interacts with the target, the photoenergy gets converted into electronic excitations immediately, and the energy relaxation through lattice takes place in ~1 picosecond (ps). Next, the photoenergy is transformed into heat diffusion (over a few microseconds (μs)), which results in the melting of the solid surface (in tens of ns). During the laser-target interactions, the localized temperature of the target reaches up to 10,000 ^{O}C or even higher, leading to the evaporation of the target material. At this point of time, the plume formation takes place (in the range of few μs). The plasma plume consists of atoms, electrons, ions and particulates of varying sizes, ranging from nanometer (nm) to micrometer (μm). This plasma reaches the substrate and undergoes re-solidification and condenses in the form of a thin film [53, 57].

In most of the cases, melting of a material depends on the rate of thermal conduction via lattice, which can be well described by the Fick's laws of diffusion. If the heated volume of the material is smaller than or equivalent to the thickness of ablated layer per laser pulse, then congruent melting will take place. Hence, PLD offers the advantage of congruent melting and vaporization. The amount of heated volume depends on the time of the laser-target interaction, i.e. the pulse duration. For a pulse duration of ~10 ps, heat diffusion will not play a role in the melting and vaporization of the material, whereas, above ~20 ps, conventional heat diffusion dependent ablation occurs [57]. Therefore, the use of a pulsed laser with a pulse duration of a few ns is more likely to provide congruent ablation. This allows the PLD process to preserve the anion-cation stoichiometry of the target material during the mass transfer of the material from the target onto the substrate.

Ambient growth pressure: The background pressure during deposition is a very important and critical parameter that plays a significant role in the plume collisions and plasma dynamics. Keeping the right background pressure is of utmost importance in order to obtain controlled stoichiometric products during the PLD growth. A specific phase and composition of a material can be achieved under controlled and optimized conditions of background pressure at specific temperature. Plasma species with kinetic energies greater than 50 eV can re-sputter the material already deposited on the substrate and this usually leads to a lower deposition rate, modifications in the stoichiometry of the film, and increase the surface roughness. Controlling the background pressure can reduce re-sputtering of the deposited thin films. Increasing the background gas pressure to an optimum value slows down the highly energetic species in the dilating plasma plume [58].

Target-substrate distance: The target to substrate distance is a useful parameter for reducing the particulate formation since majority of the PLD depositions are carried out in high pressure conditions. If the thin film is deposited in vacuum environment, the target to substrate distance mainly affects the angular spread of the ejected flux. Thus, the effect of target to substrate distance and the background gas pressures is inter-related. The plume length decreases as the ambient gas pressure increases, because of the increased collisions between the plume species and background gas molecules. Therefore, a smaller target to substrate distance should be kept in the case of depositions carried out at high pressures.

Deposition rate: This mainly depends on the repetition rate or the frequency of laser shots which controls the volume or amount of the plume species reaching the substrate and, thus, controls the thickness of the deposited thin film. Deposition rate also depends on the background gas pressure as described previously and mainly modulates the super-saturation process during deposition, which has an influence on the critical nucleation point and the density of nucleation sites. Also, ultra-smooth thin films can be obtained at optimized deposition rates.

Substrate temperature: Substrate temperature plays a critical role in terms of the diffusion barrier during the growth process and strongly affects the growth modes. It influences the nucleation process as well as the mobility of the condensed species across the substrate, and therefore, is crucial in deciding the phase boundary in the crystalline thin films during PLD growth. At lower substrate temperatures, the thin film produced may be amorphous or polycrystalline due to the lower nucleation rate, as the thermal energy provided is too small for overcoming the nucleation barrier. When the substrate temperature is too high, the nucleation rate gets limited due to the high rate of atomic exchange between the solid and gaseous species. Thus, an optimum temperature is required for the easy crystallization of thin films as it becomes easier to overcome the nucleation barrier and form nuclei on the substrate [57].

4. Recent advancements in the PLD growth of TMDCs-based PDs

In general, the PLD-grown TMDCs-based PDs exhibit device performance that is comparable with the PDs based on traditional bulk semiconductors. Additionally, PLD is also beneficial for scalable production up to the wafer-scale. Therefore, growth of these TMDCs through PLD for applications in photodetection shows a tremendous potential to translate the fundamental laboratory research to realization of industrial and practical applications.

MoS_2 is probably the most studied material among various TMDCs and was probably the first member to be fabricated by PLD. One of the earliest investigations on the photodetection studies of MoS_2-based PDs was done by Alkis *et al.* [59], in which the authors have fabricated MoS_2 nanocrystallites through PLD in deionized water and have demonstrated ultraviolet photodetection using the thin films of the obtained MoS_2 nanocrystallites. Mostly, the PLD fabricated PDs based on the TMDCs are in the form of thin films. The earliest MoS_2 thin film-based PD grown by PLD can be traced back to 2014, when Late *et al.* [60] synthesized wafer-scale MoS_2 thin films on flexible kapton substrates. The devices showed a good photoresponse towards UV light, with stable response and recovery in self-powered mode (**Figure 4a** and **b**). The origination of this self-powered behavior might be from the unintentional inhomogeneities in the contact electrodes [61, 62]. However, the observed response was very weak in the zero-biased mode. This work demonstrated that layered TMDCs can be promising candidates to be used as flexible devices in future photonic applications. In the meantime, researchers across the world started to explore the synthesis of other TMDCs by PLD. In 2015, Yao *et al.* [47] deposited multilayered WS_2 by PLD and performed detailed and systematic investigations on its photodetection properties (**Figure 4c–f**). The synthesized device exhibited a broadband and reproducible photoresponse with good stability. The photoresponse in ambient conditions reached 0.51 AW^{-1}, which was several orders higher than the CVD-grown WS_2 thin films. In vacuum conditions, the responsivity was found to be enhanced to a value of 0.7 AW^{-1}. The lower responsivity in ambient conditions has been explained on the basis of oxygen molecules adsorbed on the surface of WS_2 which trap conduction electrons, and form O_2^-. These species act as recombination centres for the photogenerated carriers. Thus, a greater

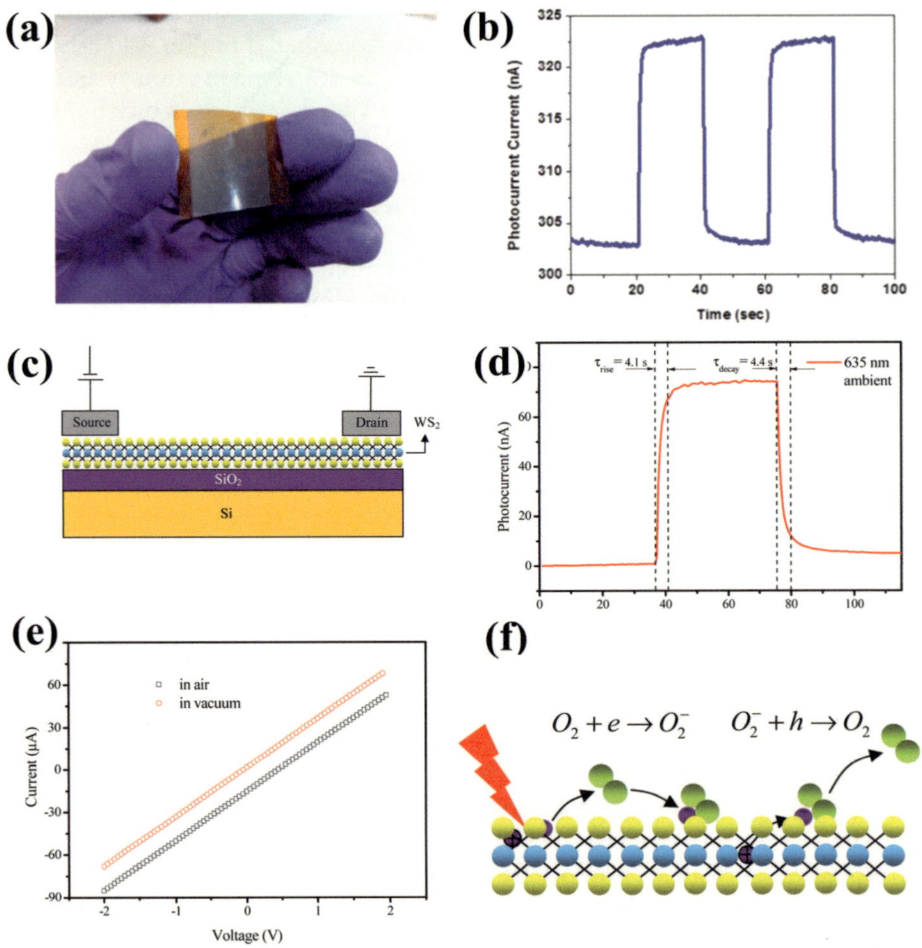

Figure 4.
(a) MoS_2 thin film deposited on flexible kapton substrate and (b) temporal response of the device in zero-bias mode. Figures have been reproduced with permission from Ref. [60]. (c) Cross-sectional schematic view of the WS_2-based photoresistor, (d) temporal response of the device in air, (e) I-V curves of the WS_2-based device in vacuum and in air, and (f) schematic of the photodetection mechanism based on adsorption and desorption of O_2 molecules. Figures have been reproduced with permission from Ref. [47].

number of adsorbed O_2 molecules are present in ambient air environment, which hampers the responsivity of the device. Furthermore, the device maintained a stable and reproducible photoswitching even after one-month of storage in air, indicating the robustness of the device.

The progress in the PLD growth of TMDCs has been quite significant, however, the crystal quality normally remains inferior when compared to the bulk natural crystals. This opens a window to further improve the device quality as well as the performance of PLD-fabricated PDs. One such work has been reported recently, where Wang *et al.* [63] have attained a dramatic improvement in the quality of the PLD-synthesized WS_2 photoresistors by using a post-synthesis annealing procedure. With increase in the post-deposition annealing temperature from 310 to 610°C, the device performance parameters of the WS_2-based PD (annealed at 610°C) enhanced by 2–3 orders of magnitude when compared to the devices annealed at lower temperatures. Annealing treatment usually provides a sufficient amount of energy and time to the atoms and molecules, for the structural reconstruction to annihilate crystal defects, and thus, has been adopted as a universal post-fabrication technique for improving the quality of the products. In another work, Yao *et al.* [64] have synthesized a hybrid $WS_2/Bi_2Te_3/SiO_2/Si$-based PD. The purpose of the insertion of the Bi_2Te_3 layer in between WS_2 and SiO_2 was to passivate the interface. The device demonstrated a stable, reproducible and broadband photoresponse (370 to 1550 nm). Moreover, the device showed a high photoresponsivity of 30.7 AW^{-1} and a pronounced specific detectivity of 2.3×10^{11} Jones with a rise time of 20 ms. The performance of the detector has been attributed to the surface passivation of SiO_2 by the Bi_2Te_3 interfacial layer. SiO_2 surface possess a lot of unscreened dangling bonds. When WS_2 is directly deposited on SiO_2, these bonds can introduce a large density of defects at the bottom of the WS_2 layer, which will act as recombination and scattering centers for the photogenerated charge carriers. With the introduction of Bi_2Te_3 layer, these dangling bonds are greatly suppressed, and this results in the growth of WS_2 film with high crystalline quality, which eventually enhances the PD's performance.

Spectral range of a PD is equally important when compared with the other figures of merit and altering the effective wavelength range of TMDC-based PDs is extremely important for specific applications. It has been shown that by introducing defect states in the forbidden gap of a semiconductor, the detection range can be dramatically extended, and sub-band gap detection can be accomplished. Xie *et al.* [65] have demonstrated ultra-broadband MoS_2-based PDs through PLD by forming sulfur vacancies in MoS_2. The S/Mo atomic ratio was modified from 1.89 to 1.94 by controlling the number of laser pulses from 1200 to 300, resulting in a dramatic increase in the band gap of the semiconductor. Consequently, the S-deficient MoS_2-based PD exhibited an unprecedented ultra-broadband detection range from 445 to 2717 nm. However, theoretical calculations have been done which indicated that the Mo vacancies in MoS_2 possess a higher capability for narrowing the band gap. Therefore, in a subsequent work, Xie *et al.* [66] have synthesized a series of Mo-deficient MoS_2-based PDs by moderating the target to substrate distance during the PLD growth. As a result of these modulations, the effective spectral range of a $MoS_{2.17}$ PD spanned all over from 445 nm to 9536 nm. Although the detection range can be extended up to mid infrared (MIR) with the introduction of the defect states, however, it is accompanied by certain challenges that hinders the usability of this method in practical devices. These include the control over the creation of these defects, which is still an unresolved problem. Furthermore, the electronic properties of the charge carriers are severely hampered owing to the increased scattering effects from these defect states. Thus, such devices often suffer from meager responsivities and slower response speeds.

Recently, Jiao *et al.* [67] have synthesized high-quality and wafer-scale 2D layered MoS_2 thin films by PLD. The device exhibited competitive device performance to the commercial Si and Ge-based PDs. The value of the responsivity was recorded to be 1.96 AW^{-1} for single layer MoS_2-based device, under 300 nm light illumination (**Figure 5**). The PD shows a broadband photoresponse ranging from UV to NIR, with a fast response of 96 ms. This enhancement in the performance was attributed to the variation in the Schottky barrier at the Au/MoS_2 interface.

The above discussed PLD-grown TMDC-based PDs are based on the metal-semiconductor-metal type device configuration, and suffer from relatively lower photoresponsivity, low on/off ratios, narrowband detection and slower detection speed. Hence, strategies are being developed to overcome these limitations. PDs having transparent electrodes such as indium tin oxide (ITO) and graphene, instead of the conventional metal contacts and PDs based on heterojunctions of two or more materials have many advantages such as low value of dark current, higher on/off current ratios, broadband detection range, and higher responsivities due to favorable band alignments.

One such work was carried out in 2016, when Zheng *et al.* [68] successfully prepared centimeter-scale and highly-crystalline WSe_2 thin films on polyimide substrates by the technique of PLD and have fabricated high-performance PDs based on these WSe_2 thin films. They obtained a broadband spectral response, ranging from 370 to 1064 nm. Moreover, a reproducible photoresponsivity approaching up to 0.92 AW^{-1}, an EQE of 180% and a fast response speed of 0.9 s have also been achieved. The PD also exhibited excellent air durability and mechanical flexibility. The enhanced performance has been attributed to the good Ohmic contacts WSe_2 forms with ITO, because of a low mismatch between the work functions of the two materials. Due to the Ohmic contacts, the carriers can be efficiently injected through the ITO electrodes under an applied bias, which will result in generation of a high photocurrent. Ohmic contacts lead to photodetection mechanisms based on the intrinsic properties of the photosensitive material under light irradiation.

Using a similar approach of integration of ITO electrodes on the device, Kumar *et al.* [69] have reported a UV PD which utilizes few layered MoS_2 deposited by PLD. The device shows a high responsivity of 3×10^4 AW^{-1} and detectivity of 1.81×10^{14} Jones, at a nominal voltage of 2 V with fast response time of 32 ms. This performance is better than most of the reported devices based on 2D layered materials. The PD exhibited a very low value of dark current (~10^{-10} A) which is the reason behind such an excellent device performance. This may be

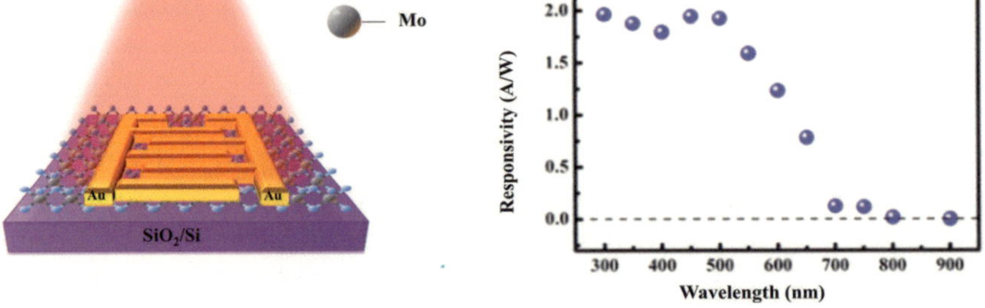

Figure 5.
(a) Schematic of the interdigital patterned gold electrodes to form a metal-semiconductor-metal type contact, and (b) the wavelength dependence of responsivity (300 to 900 nm) for the device. Figures have been reproduced with permission from Ref. [67].

because of suitable band alignment with the ITO electrodes as well as the deposition of high-quality films as the deposition was carried out in the presence of nitrogen gas, leading to lower number of sulfur vacancies.

Till now, all the above reported devices require some external bias for obtaining significant photodetection. Over the years, PDs which consume no external power have attracted a lot of attention because in the current scenario of energy crisis, a lot of research has been focused on energy producing and energy storage devices [70–74]. Such self-powered PDs depend on the interfacial built-in potential which enhances the effective separation of photogenerated carriers. This built-in electric potential also suppresses the dark current, which is another benefit for such PDs. Therefore, these self-powered devices have a great prospect for the next-generation PDs.

In 2015, Yao *et al*. [75] designed a Bi/WS_2/Si heterojunction-based PD by depositing polycrystalline WS_2 and Bi thin films on a p-type Si substrate by PLD (**Figure 6**). The PD exhibited a decent responsivity of 0.42 AW^{-1} and a high detectivity of 1.36×10^{13} Jones with ultrahigh sensitivity. It was observed that the performance of the Bi/Si heterointerface was enhanced by the insertion of the WS_2 film. The enhanced device performance has been attributed to the effective passivation of the junction, and enhanced light absorption. Moreover, due to the favorable band alignment, WS_2 acts as a selective carrier blocker, which further enhances the device performance.

Recently, Singh *et al*. [4] demonstrated an MoS_2/AlN/Si-based PD grown by PLD, thus combining the excellent and unique properties of MoS_2 with the matured technologies of Si and III-nitride semiconductors. Moreover, due to a large

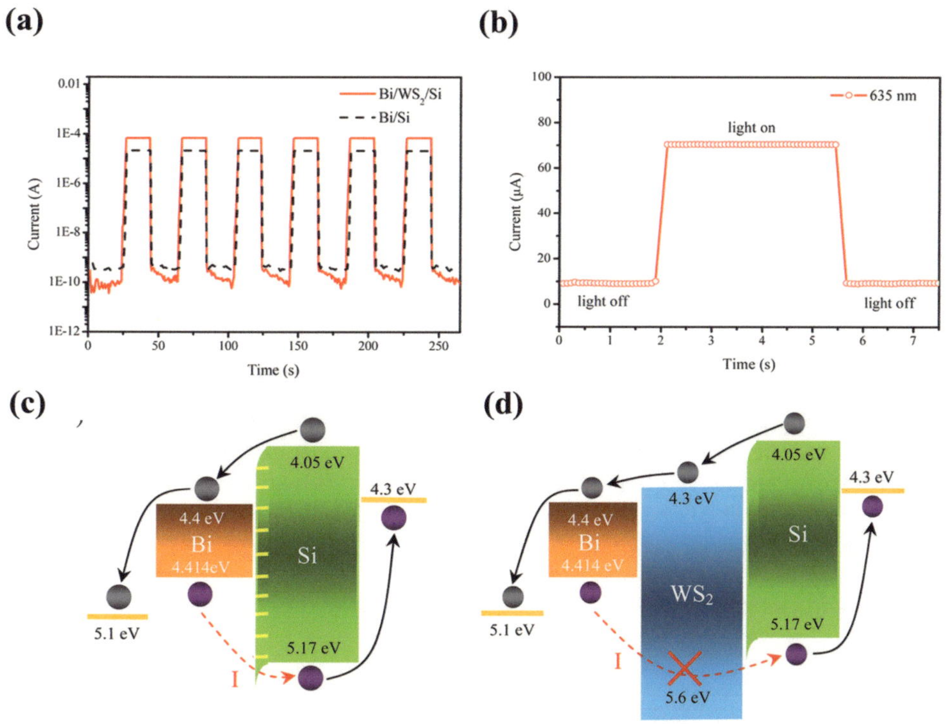

Figure 6.
(a) Transient behavior of the Bi/Si and Bi/WS_2/Si PDs under zero bias. (b) Corresponding single on-off cycle. Schematic of the energy band diagrams of the (c) Bi/Si and (d) Bi/WS_2/Si heterointerfaces. Figures have been reproduced with permission from Ref. [75].

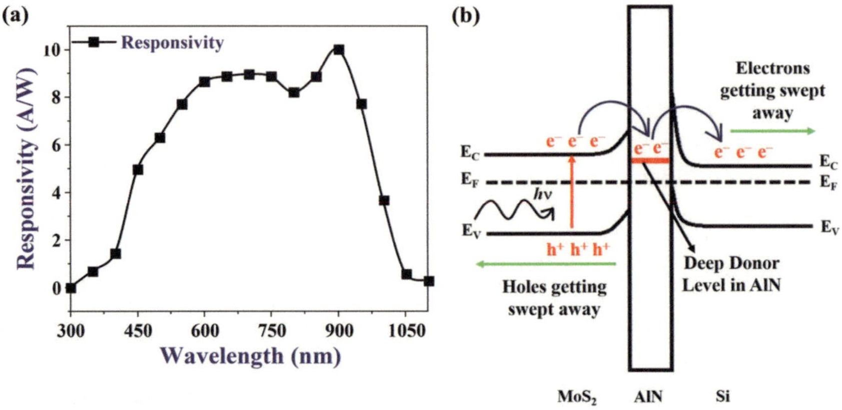

Figure 7.
(a) Spectral response of the MoS_2/AlN/Si-based PD. (b) Schematic of the deep defect states-modulated carrier transport in MoS_2/AlN/Si-based device. Figures have been reproduced with permission from Ref. [4].

difference in the work functions of these materials, the band bending of the heterojunction at the interfaces resulted in self-powered behavior. The vertical transport of the PD exhibited an exceptional broadband photoresponse (300–1100 nm) in the self-powered mode. The device shows a responsivity of 9.93 AW^{-1} under zero-bias condition with ultrafast response speeds (response/recovery times - 12.5/14.9 μs). The photoresponse of MoS_2/Si has also been given, to show the importance of inserting the AlN layer. The MoS_2/Si PD exhibits a responsivity of 1.88 AW^{-1} (~5 times less) in self-powered mode. The authors have shown that the native oxygen defects are present throughout the AlN layer, and this has been confirmed with the help of X-ray photoelectron spectroscopy and transmission electron microscopy. These oxygen impurities form deep donor states in AlN and modulate the transport of the charge carriers, and this leads to the enhanced performance of the MoS_2/AlN/Si-based device (**Figure 7**).

5. Summary

The past few years have undoubtedly witnessed tremendous advances in the PLD growth of TMDCs and their applications in the field of photodetection. In this chapter, the basic properties of TMDCs and the common growth techniques employed for their fabrication have been reviewed briefly, followed by a detailed and elaborated discussion about PLD and the important parameters associated with it. Finally, a progressive investigation about the PDs based on TMDCs fabricated through PLD has been discussed. These extensive achievements in the field of photodetection have unquestionably established PLD as one of the most competitive and reliable methods for fabricating industrial-scale and high-quality TMDCs. PLD, therefore, certainly has a lot of potential to contribute in the development of the next-generation TMDCs-based PDs in the future.

6. Looking into the future

Based on the analysis of the previous reports in this area, an outlook regarding the future of PLD-synthesized TMDCs and the related follow-up research work has been summarized below.

- The research on the PLD-fabricated TMDCs-based PDs is still in its nascent stage and therefore, there is still a room for improvement in the crystal-quality of the PLD-grown thin films, by selecting appropriate substrates and by further tuning and optimizing the various unexplored growth parameters, such as annealing temperature and time, cooling ramp rate, geometry of the laser spot, surface morphology of the source targets, laser frequency, and so on [18].

- Apart from a few reports, most of the research regarding PLD growth of TMDCs for photodetection application is based on the use of a single photosensitive material. Therefore, promising results are expected on the exploration of the heterojunctions of these layered materials with established bulk semiconductors like III-nitrides, which have shown great results in this field. Moreover, TMDCs can serve as excellent substrates for high quality and epitaxial growth of III-nitrides, which would lead to better device performance.

- The use of transparent 2D semiconductors such as graphene or graphene derivatives and semi-metallic phase of TMDCs, can be used in place of the conventional metal electrodes, as they maximize the area of light absorption along with having outstanding electronic properties.

- Till date, researchers and scientists across the world have mostly exploited heterojunctions of TMDCs in their thin film forms. Heterointerfaces based on one dimensional (1D) nanostructures may provide new routes for the development of high-performance devices. The nanowire-based heterostructures of the TMDCs with growth along these nanowires or a core-shell structure will enable a much higher surface to volume ratio and therefore, a larger active interface. This will lead to enhancement in the photoresponse and superior optoelectronic performance.

As a concluding statement, PLD has been proven to be a promising synthesis technique for TMDCs for applications in photodetection, and these PDs have shown outstanding performance that can compete with those of the commercially available PDs. The fabrication through PLD is cost-effective and scalable, and hence, PLD is a perfect tool for fabrication of practical devices for optoelectronic applications at an industrial scale production.

Acknowledgements

D.K.S. is thankful to Council of Scientific and Industrial Research, Government of India, New Delhi for providing senior research fellowship. S.B.K. acknowledges INSA senior scientist fellowship.

Author details

Deependra Kumar Singh, Karuna Kar Nanda and Saluru Baba Krupanidhi*
Materials Research Centre, Indian Institute of Science, Bangalore, India

*Address all correspondence to: sbkrupanidhi@gmail.com

References

[1] Wu W, Zhang Q, Zhou X, Li L, Su J, Wang F, Zhai T. Self-powered photovoltaic photodetector established on lateral monolayer MoS_2-WS_2 heterostructures. Nano Energy. 2018;51: 45-53.

[2] Singh RK, Kumar J, Kumar A, Kumar V, Kant R, Singh R. Poly(3-hexylthiophene): Functionalized single-walled carbon nanotubes: (6,6)-phenyl-C61-butyric acid methyl ester composites for photovoltaic cell at ambient condition. Solar Energy Materials and Solar Cells. 2010;94(12):2386-94.

[3] Khan MA, Nanda KK, Krupanidhi SB. Mechanistic view on efficient photodetection by solvothermally reduced graphene oxide. Journal of Material Science: Material in Electronics. 2017;28(19):14818-26.

[4] Singh DK, Pant R, Chowdhury AM, Roul B, Nanda KK, Krupanidhi SB. Defect-Mediated Transport in Self-Powered, Broadband, and Ultrafast Photoresponse of a MoS_2/AlN/Si-Based Photodetector. ACS Applied Electronic Materials. 2020;2(4):944-53.

[5] Lopez-Sanchez O, Lembke D, Kayci M, Radenovic A, Kis A. Ultrasensitive photodetectors based on monolayer MoS_2. Nature Nanotechnology. 2013;8(7):497-501.

[6] Pant RK, Singh DK, Roul B, Chowdhury AM, Chandan G, Nanda KK, et al. Photodetection Properties of Nonpolar a-Plane GaN Grown by Three Approaches Using Plasma-Assisted Molecular Beam Epitaxy. physica status solidi (a). 2019;216(18):1900171.

[7] Sai Manohar GV, Krupanidhi SB, Nanda KK. Giant enhancement in photoresponse via engineering of photo-induced charge (electron and hole) transfer in linear and non-linear devices. Sensors and Actuators A: Physical. 2020;304:111842.

[8] Chowdhury AM, Chandan G, Pant R, Roul B, Singh DK, Nanda KK, et al. Self-Powered, Broad Band, and Ultrafast InGaN-Based Photodetector. ACS Applied Material & Interfaces. 2019;11(10):10418-25.

[9] Hsu L-H, Kuo C-T, Huang J-K, Hsu S-C, Lee H-Y, Kuo H-C, et al. InN-based heterojunction photodetector with extended infrared response. Optical Express. 2015;23(24):31150.

[10] Arora K, Kumar M. Sputtered-Growth of High-Temperature Seed-Layer Assisted β-Ga_2O_3 Thin Film on Silicon-Substrate for Cost-Effective Solar-Blind Photodetector Application. ECS Journal of Solid State Science and Technology. 2020;9(6):065013.

[11] Arora K, Singh DP, Fischer P, Kumar M. Spectrally Selective and Highly Sensitive UV Photodetection with UV-A,C Band Specific Polarity Switching in Silver Plasmonic Nanoparticle Enhanced Gallium Oxide Thin-Film. Advanced Optical Materials. 2020;8(16):2000212.

[12] Peytavit E, Arscott S, Lippens D, Mouret G, Matton S, Masselin P, et al. Terahertz frequency difference from vertically integrated low-temperature-grown GaAs photodetector. Applied Physics Letters. 2002;81(7):1174-6.

[13] Berencén Y, Prucnal S, Liu F, Skorupa I, Hübner R, Rebohle L, et al. Room-temperature short-wavelength infrared Si photodetector. Scientific Reports. 2017;7(1):43688.

[14] Monroy E, Calle F, Pau JL, Muñoz E, Omnès F, Beaumont B, et al. AlGaN-based UV photodetectors. Journal of Crystal Growth. 2001;230(3-4):537-43.

[15] Masini G, Cencelli V, Colace L, DeNotaristefani F, Assanto G. A germanium photodetector array for the near infrared monolithically integrated with silicon CMOS readout electronics. Physica E: Low-dimensional Systems and Nanostructures. 2003;16(3-4):614-9.

[16] Zhou YD, Becker CR, Selamet Y, Chang Y, Ashokan R, Boreiko RT, et al. Far-infrared detector based on HgTe/HgCdTe superlattices. Journal of Electronic Materials. 2003;32(7):608-14.

[17] Bowers JE, Srivastava AK, Burrus CA, DeWinter MA, Pollack MA, Zyskind JL. High-speed GaInAsSb/GaSb PIN photodetectors for wavelengths to 2.3 μm. Electronics Letters. 1986;22(3):137-8.

[18] Yao JD, Zheng ZQ, Yang GW. Production of large-area 2D materials for high-performance photodetectors by pulsed-laser deposition. Progress in Materials Science. 2019;106:100573.

[19] Novoselov KS, Geim AK, Morozov SV, Jiang D, Zhang Y, Dubonos SV, et al. Electric field effect in atomically thin carbon films. Science. 2004;306(5696):666-9.

[20] Thakur MK, Gupta A, Ghosh S, Chattopadhyay S. Graphene-Conjugated Upconversion Nanoparticles as Fluorescence-Tuned Photothermal Nanoheaters for Desalination. ACS Applied Nano Materials. 2019;2(4):2250-9.

[21] Thakur MK, Fang C-Y, Yang Y-T, Effendi TA, Roy PK, Chen R-S, et al. Microplasma-Enabled Graphene Quantum Dot-Wrapped Gold Nanoparticles with Synergistic Enhancement for Broad Band Photodetection. ACS Applied Mater & Interfaces. 2020;12(25):28550-60.

[22] Thakur MK, Gupta A, Fakhri MY, Chen RS, Wu CT, Lin KH, et al. Optically coupled engineered upconversion nanoparticles and graphene for a high responsivity broadband photodetector. Nanoscale. 2019;11(19):9716-25.

[23] Liu HF, Wong SL, Chi DZ. CVD Growth of MoS_2-based Two-dimensional Materials. Chemical Vapor Deposition. 2015;21(10-11-12):241-59.

[24] Gonzalez JM, Oleynik II. Layer-dependent properties of SnS_2 and $SnSe_2$ two-dimensional materials. Physical Review B. 2016;94(12):125443.

[25] Lin Y-M, Dimitrakopoulos C, Jenkins KA, Farmer DB, Chiu H-Y, Grill A, et al. 100-GHz Transistors from Wafer-Scale Epitaxial Graphene. Science. 2010;327(5966):662-662.

[26] Li C, Yan X, Song X, Bao W, Ding S, Zhang DW, et al. WSe_2/MoS_2 and $MoTe_2/SnSe_2$ van der Waals heterostructure transistors with different band alignment. Nanotechnology. 2017;28(41):415201.

[27] Nam S-G, Cho Y, Lee M-H, Shin KW, Kim C, Yang K, et al. Barrier height control in metal/silicon contacts with atomically thin MoS_2 and WS_2 interfacial layers. 2D Materials. 2018;5(4):041004.

[28] Ko PJ, Abderrahmane A, Kim N, Sandhu A. High-performance near-infrared photodetector based on nano-layered $MoSe_2$. Semiconductor Science and Technology. 2017;32(6):065015.

[29] Yin Z, Li H, Li H, Jiang L, Shi Y, Sun Y, et al. Single-Layer MoS_2 Phototransistors. ACS Nano. 2012;6(1):74-80.

[30] Zhuo R, Wang Y, Wu D, Lou Z, Shi Z, Xu T, et al. High-performance self-powered deep ultraviolet photodetector based on MoS_2/GaN p–n heterojunction. Journal of Materials Chemistry C. 2018;6(2):299-303.

[31] Kang M-A, Kim S, Jeon I-S, Lim YR, Park C-Y, Song W, et al. Highly efficient and flexible photodetector based on MoS_2–ZnO heterostructures. RSC Advances. 2019;9(34):19707-11.

[32] Siegel G, Venkata Subbaiah YP, Prestgard MC, Tiwari A. Growth of centimeter-scale atomically thin MoS $_2$ films by pulsed laser deposition. APL Materials. 2015;3(5):056103.

[33] Manzeli S, Ovchinnikov D, Pasquier D, Yazyev OV, Kis A. 2D transition metal dichalcogenides. Nature Reviews Materials. 2017;2(8):17033.

[34] Splendiani A, Sun L, Zhang Y, Li T, Kim J, Chim C-Y, et al. Emerging Photoluminescence in Monolayer MoS_2. Nano Letters. 2010;10(4):1271-5.

[35] Singh A, Singh AK. Origin of n -type conductivity of monolayer MoS_2. Physical Review B. 2019;99(12):121201.

[36] Tangi M, Mishra P, Ng TK, Hedhili MN, Janjua B, Alias MS, et al. Determination of band offsets at GaN/single-layer MoS_2 heterojunction. Applied Physics Letters. 2016;109(3):032104.

[37] Pulsed Laser Deposition (PLD). https://vaccoat.com/blog/pulsed-laser-deposition-pld/.

[38] Serna MI, Yoo SH, Moreno S, Xi Y, Oviedo JP, Choi H, et al. Large-Area Deposition of MoS_2 by Pulsed Laser Deposition with *In Situ* Thickness Control. ACS Nano. 2016;10(6):6054-61.

[39] Singh DK, Roul B, Pant R, Chowdhury AM, Nanda KK, Krupanidhi SB. Different types of band alignment at MoS_2/(Al, Ga, In) N heterointerfaces. Appl Phys Lett. 2020;116(25):252102.

[40] Zabinski JS, Donley MS, Dyhouse VJ, McDevitt NT. Chemical and tribological characterization of PbO:MoS_2 films grown by pulsed laser deposition. Thin Solid Films. 1992;214(2):156-63.

[41] McDevitt NT, Bultman JE, Zabinski JS. Study of Amorphous MoS_2 Films Grown by Pulsed Laser Deposition. Applied spectroscopy. 1998;52(9):1160-4.

[42] Mosleh M, Laube SJP, Suh NP. Friction of Undulated Surfaces Coated with MoS_2 by Pulsed Laser Deposition. Tribology Transactions. 1999;42(3):495-502.

[43] Fominski VYu, Romanov RI, Gnedovets AG, Nevolin VN. Formation of the chemical composition of transition metal dichalcogenide thin films at pulsed laser deposition. Technical Physics. 2010;55(10):1509-16.

[44] Loh TAJ, Chua DHC. Growth Mechanism of Pulsed Laser Fabricated Few-Layer MoS_2 on Metal Substrates. ACS Applied Materials & Interfaces. 2014;6(18):15966-71.

[45] Serrao CR, Diamond AM, Hsu S-L, You L, Gadgil S, Clarkson J, et al. Highly crystalline MoS_2 thin films grown by pulsed laser deposition. Applied Physics Letters. 2015;106(5):052101.

[46] Loh TAJ, Chua DHC, Wee ATS. One-step Synthesis of Few-layer WS_2 by Pulsed Laser Deposition. Scientific Reports. 2016;5(1):18116.

[47] Yao JD, Zheng ZQ, Shao JM, Yang GW. Stable, highly-responsive and broadband photodetection based on large-area multilayered WS_2 films grown by pulsed-laser deposition. Nanoscale. 2015;7(36):14974-81.

[48] Ullah F, Nguyen TK, Le CT, Kim YS. Pulsed laser deposition assisted grown continuous monolayer $MoSe_2$. CrystEngComm. 2016;18(37):6992-6.

[49] Mohammed A, Nakamura H, Wochner P, Ibrahimkutty S, Schulz A,

Müller K, et al. Pulsed laser deposition for the synthesis of monolayer WSe_2. Applied Physics Letters. 2017;111(7):073101.

[50] Seo S, Choi H, Kim SY, Lee J, Kim K, Yoon S, et al. Growth of Centimeter-Scale Monolayer and Few-Layer WSe_2 Thin Films on SiO_2/Si Substrate via Pulsed Laser Deposition. Advanced Materials Interfaces. 2018;5(20):1800524.

[51] Gao M, Zhang M, Niu W, Chen Y, Gu M, Wang H, et al. Tuning the transport behavior of centimeter-scale WTe_2 ultrathin films fabricated by pulsed laser deposition. Applied Physics Letters. 2017;111(3):031906.

[52] Maiman TH. Ruby laser systems. United States patent, US 3,353,115. 1967.

[53] Stafe M, Marcu A, Puscas NN. Pulsed Laser Ablation of Solids - Basics, Theory and Applications. Springer Series in Surface Sciences; 2014.

[54] Eason R. Pulsed laser deposition of thin films: applications-led growth of functional materials. John Wiley & Sons; 2007.

[55] Singh RK, Narayan J. A novel method for simulating laser-solid interactions in semiconductors and layered structures. Materials Science and Engineering: B. 1989;3(3):217-30.

[56] Wood RF, Giles GE. Macroscopic theory of pulsed-laser annealing. I. Thermal transport and melting. Physical Review B. 1981;23(6):2923-42.

[57] Daramalla, V. Titanium Niobium complex oxide ($TiNb_2O_7$) thin films for micro battery applications. PhD thesis. Indian Institute of Science, Bangalore; 2015.

[58] Gai Z, Kalinin SV, Li A-P, Shen J, Baddorf AP. In Situ Observations and Tuning of Physical and Chemical Phenomena on the Surfaces of Strongly Correlated Oxides. Advanced Functional Materials. 2013;23(20):2477-89.

[59] Alkis S, Öztaş T, Aygün LE, Bozkurt F, Okyay AK, Ortaç B. Thin film MoS_2 nanocrystal based ultraviolet photodetector. Optics Express. 2012;20(19):21815.

[60] Late DJ, Shaikh PA, Khare R, Kashid RV, Chaudhary M, More MA, et al. Pulsed Laser-Deposited MoS_2 Thin Films on W and Si: Field Emission and Photoresponse Studies. ACS Applied Materials & Interfaces. 2014;6(18):15881-8.

[61] Pant R, Singh DK, Chowdhury AM, Roul B, Nanda KK, Krupanidhi SB. Highly Responsive, Self-Powered *a*-GaN Based UV-A Photodetectors Driven by Unintentional Asymmetrical Electrodes. ACS Applied Electronic Materials. 2020;2(3):769-79.

[62] Rambabu A, Singh DK, Pant R, Nanda KK, Krupanidhi SB. Self-powered, ultrasensitive, room temperature humidity sensors using SnS_2 nanofilms. Scientific Reports. 2020;10(1):14611.

[63] Wang H, Ng SM, Wong HF, Wong WC, Lam KK, Liu YK, et al. Effect of post-annealing on laser-ablation deposited WS_2 thin films. Vacuum. 2018;152:239-42.

[64] Yao J, Zheng Z, Yang G. Layered-material WS_2/topological insulator Bi_2Te_3 heterostructure photodetector with ultrahigh responsivity in the range from 370 to 1550 nm. Journal of Materials Chemistry C. 2016;4(33):7831-40.

[65] Xie Y, Zhang B, Wang S, Wang D, Wang A, Wang Z, et al. Ultrabroadband MoS_2 Photodetector with Spectral Response from 445 to 2717 nm. Advanced Materials. 2017;29(17):1605972.

[66] Xie Y, Liang F, Wang D, Chi S, Yu H, Lin Z, et al. Room-Temperature Ultrabroadband Photodetection with MoS_2 by Electronic-Structure Engineering Strategy. Advanced Materials. 2018;30(50):1804858.

[67] Jiao L, Jie W, Yang Z, Wang Y, Chen Z, Zhang X, et al. Layer-dependent photoresponse of 2D MoS_2 films prepared by pulsed laser deposition. Journal of Materials Chemistry C. 2019;7(9):2522-9.

[68] Zheng Z, Zhang T, Yao J, Zhang Y, Xu J, Yang G. Flexible, transparent and ultra-broadband photodetector based on large-area WSe_2 film for wearable devices. Nanotechnology. 2016;27(22):225501.

[69] Kumar S, Sharma A, Ho YT, Pandey A, Tomar M, Kapoor AK, et al. High performance UV photodetector based on MoS_2 layers grown by pulsed laser deposition technique. Journal of Alloys and Compounds. 2020;835:155222.

[70] Shi D, Wang G, Li C, Shen X, Nie Q. Preparation and thermoelectric properties of $MoTe_2$ thin films by magnetron co-sputtering. Vacuum. 2017;138:101-4.

[71] Kumar A, Singh RK, Singh HK, Srivastava P, Singh R. Enhanced capacitance and stability of p-toluenesulfonate doped polypyrrole/carbon composite for electrode application in electrochemical capacitors. Journal of Power Sources. 2014;246:800-7.

[72] Kumar S, Chaudhary D, Khare N. Enhanced thermoelectric figure of merit in Bi_2Te_3–CNT–PEDOT nanocomposite by introducing conducting interfaces in Bi_2Te_3 nanostructures. APL Materials. 2019;7(8):081123.

[73] Kumar A, Singh RK, Singh HK, Srivastava P, Singh R. Mechanism of direct current electrical charge conduction in *p*-toluenesulfonate doped polypyrrole/carbon composites. Journal of Applied Physics. 2014;115(10):103702.

[74] Kumar A, Singh HK, Singh RK, Singh R, Srivastava P. P-toluenesulfonate doped polypyrrole/carbon composite electrode and a process for the preparation thereof. United States patent US 10,074,453. 2018.

[75] Yao J, Zheng Z, Shao J, Yang G. Promoting Photosensitivity and Detectivity of the Bi/Si Heterojunction Photodetector by Inserting a WS_2 Layer. ACS Applied Materials & Interfaces. 2015;7(48):26701-8.

Chapter 3

Binary Metal Oxides Thin Films Prepared from Pulsed Laser Deposition

Cyril Robinson Azariah John Chelliah and Rajesh Swaminathan

Abstract

The semiconductor industry flourished from a simple Si-based metal oxide semiconductor field effect transistor to an era of MOSFET-based smart materials. In recent decades, researchers have been replacing all the materials required for the MOSFET device. They replaced the substrate with durable materials, lightweight materials, translucent materials and so on. They have came up with the possibility of replacing dielectric silicon dioxide material with high-grade dielectric materials. Even then the channel shift in the MOSFET was the new trend in MOSFET science. From the bulk to the atomic level, transistors have been curiously researched across the globe for the use of electronic devices. This research was also inspired by the different semiconductor materials relevant to the replacement of the dielectric channel/gate. Study focuses on diverse materials such as zinc oxides (ZnO), electrochromic oxides such as molybdenum oxides (including MoO_3 and MoO_2) and other binary oxides using ZnO and MoO_3. The primary objective of this research is to study pulsed laser deposited thin films such as ZnO, MoO_3, binary oxides such as binary ZnO /MoO_3, ZnO /TiO_2 and ZnO/V_2O_5 and to analyse their IV properties for FET applications. To achieve the goal, the following working elements have been set: investigation of pulsed laser deposited thin film of metal oxides and thin film of binary metal oxide nanostructures with effects of laser repetition and deposition temperatures.

Keywords: binary oxides, PLD, thin film, metal oxides

1. Introduction

Metal oxide thin film plays an important role in various applications such as aircraft cockpits, electronic displays, medical devices, solar panels, smart windows and high-temperature sensors in spacecraft, photonics, photodetectors, infrared detectors, phototransistors, transparent electronics, optics, anti-reflective and decorative coatings [1]. They own excellent banding and carrier mobility for conduction phenomena in the field effect transistor application. The semiconductor industry flourished from a simple Si-based metal oxide semiconductor field effect transistor to an era of MOSFET-based smart materials. In recent decades, researchers have been replacing all the materials needed for the MOSFET unit. They replaced the substrate with flexible materials, light weight material, transparent material, etc. They also come up with the idea of replacing dielectric silicon dioxide material with high-grade dielectric materials. And changing the channel in the MOSFET has become the latest trend in FET research. From the bulk to the atomic level, transistors have been interestingly studied around the globe for the application of electronic devices. This research was therefore motivated by the various semiconductor materials applicable to the replacement of the dielectric channel/gate [2, 3].

Research focuses on various materials such as zinc oxides (ZnO), electrochromic oxides such as molybdenum oxides (including MoO_3 and MoO_2) and various binary oxides using ZnO and MoO_3. New electrical results are studied with the investigation of the conduction mechanism using ac complex impedance spectroscopy with ZnO and MoO_3 binary oxides. This has motivated the research to work on new binary materials with ZnO /TiO_2 and ZnO /V_2O_5. As a result, we investigated the various electrical properties and complex impedance parameters using ac impedance spectroscopy. An attempt with pulsed laser deposited ZnO thin film as a channel layer and Al_2O_3 as a dielectric layer in FET was fabricated and studied. A transistor with ZnO channel, MoO_3 as an interlayer with source-substrate and drain-substrate, and a layer of binary oxides ZnO/MoO_3 and ZrO_2 as a dielectric stack layer is also manufactured and analysed.

The primary objective of this research is to study the pulsed laser deposited thin films such as ZnO, MoO_3, binary oxides such as ZnO/MoO_3, ZnO/TiO_2 and ZnO/V_2O_5 and their behaviour as gate dielectric layer to investigate the IV properties for FET applications. To achieve the objective following work elements were set: First to investigate the pulsed laser deposited thin films of ZnO nanostructures with the effects of laser repetition rate and deposition temperatures, then an investigation of the MoO_3 and MoO_2 thin films by pulsed laser deposition with the impact of O_2 and Ar atmosphere gas and deposition temperatures. The research continued its next set of investigation of the current conduction mechanism of the pulsed laser deposited binary oxide ZnO/MoO_3, ZnO/TiO_2 and ZnO/V_2O_5 thin films using ac complex impedance spectroscopy and the impact of wide range of temperature from 298 K to 423 K and a wide range of frequency from 1 Hz to 1 MHz. Finally, a ZnO thin film by pulsed laser deposition as n-channel MOSFET and its performance of ZnO channel along with various binary oxides and interlayer MoO_3 in MOSFET were studied [4].

2. Materials and methods

2.1 General information on zinc oxide

Zinc oxide is a unique material that exhibits exceptional semiconducting, piezoelectric, and pyroelectric properties. Nanostructures of ZnO are equally as important as carbon nanotubes and silicon nanowires for nanotechnology and have great potential applications in nanoelectronics, optoelectronics, sensors, field emission, light-emitting diodes, photocatalysis, nanogenerators, and nanopiezoelectronics. Fundamental understanding about the growth of ZnO nanowires is of critical importance for controlling their size, composition, structure, and corresponding physical and chemical properties [5].

2.2 General information on transition metal oxide

Moreover, the MoO_3, V_2O_5 and TiO_2 belong to the transition metal oxides (TMO) family. Hence these mixed metal oxides MoO_3, V_2O_5 and TiO_2 also called as binary transition metal oxides (BTMOs). They consist of at least one transition metal ion and one or more electrochemically active/inactive ions. They can use the synergism behavior of pure oxides, which can enrich the capacitive performance with an expanded potential window, supplementary active sites, excellent conductivity and improved stability. According to Zhang et al. [6], BTMOs possess higher reversible capacity, better structural stability and electronic conductivity, and have been widely studied to be novel electrode materials for transistor applications.

2.3 Basic properties of ZnO, MoO_3, TiO_2 and V_2O_5

Table 1 provides the basic properties of ZnO, MoO_3, TiO_2 and V_2O_5.

2.4 Pulsed laser deposition

The various thin film deposition methods are sol–gel deposition, thermal electrodeposition, chemical vapor deposition, sputtering, atomic layer deposition, pulsed laser deposition and so on. Among these the pulsed laser deposition is a sophisticated, rich and excellent deposition method especially for metal oxides (MO) thin films depositions.

2.5 Physical principle

A pulsed laser beam is allowed to hit the target material which is placed on the target holder. This is fixed with the rotating target carrousel. A laser plume consisting of atoms, molecules, ions, nanoparticles and microparticles also known as plasma produced. This plasma of molten material by the pulsed laser starts

Properties	Zinc oxide	Molybdenum oxide	Titanium dioxide	Vanadium pentoxide
Molecular formula	ZnO	MoO_3	TiO_2	V2O5
Molar mass (g/mol)	81.39	143.94	79.87	149.88
Appearance	White solid	Colorless or white to slightly bluish powder or granules	White powder	Yellow to red crystalline powder
Odor	Odorless	Odorless	Odorless	Odorless
Density (g/cm^3)	5.61	4.70	4.23	3.36
Refractive index	2.0041	2.5166	2.9103	2.3
Melting point	2248 K (decomposes)	1068 K	2128 K	2213 K
Boiling point	2633 K	1428 K (sublimes)	3273 K	2023 K (decomposes)
Solubility in water	0.16 mg/100 mL (at 30 °C) (Insoluble in water)	0.49 g/1000 mL in water at 28 °C	Insoluble in water	1 g / 125 mL (slightly soluble)
Bandgap (eV)	3.34 (direct)	2.9 (direct)	3.2 (indirect)	2.8 (indirect)
Dielectric constant	8.33	35	80	37.2
CAS number	1314-13-2	1313-27-5	13463–67-7	1314-62-1

Table 1.
Basic properties of the ZnO, MoO_3, TiO_2 and V_2O_5.

depositing on the substrate a rich, vibrant, stoichiometric and neatly arranged self-assembled layers. **Figure 1** depicts the schematic of the working principle of the PLD.

2.6 Working operation of PLD

Figure 2(a) and **(b)** shows the flowchart of the experiments to be carried out in a standard procedure before and after the deposition process using the pulsed laser deposition system.

2.7 Advantages of using PLD

The advantages of using PLD include:

i. wavelength and power density flexibility help to ablate any material combination.

ii. laser system is separated from the vacuum system hence while in the ablation geometry there is a considerable degree of freedom

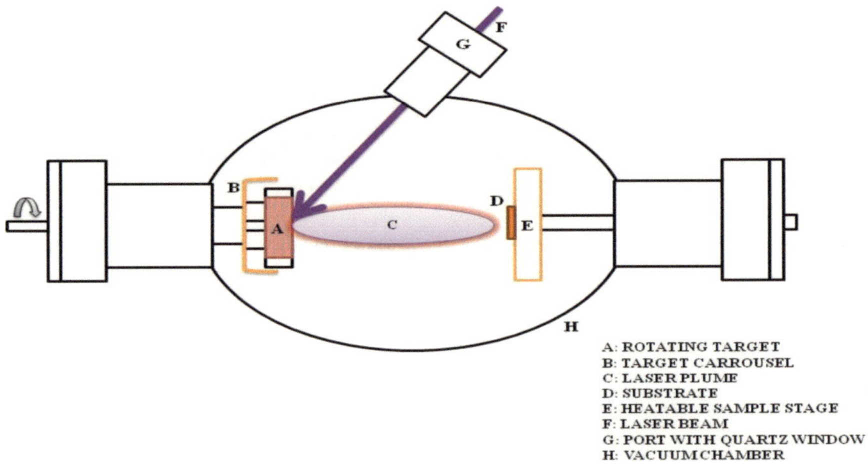

Figure 1.
Schematic working of pulsed laser deposition.

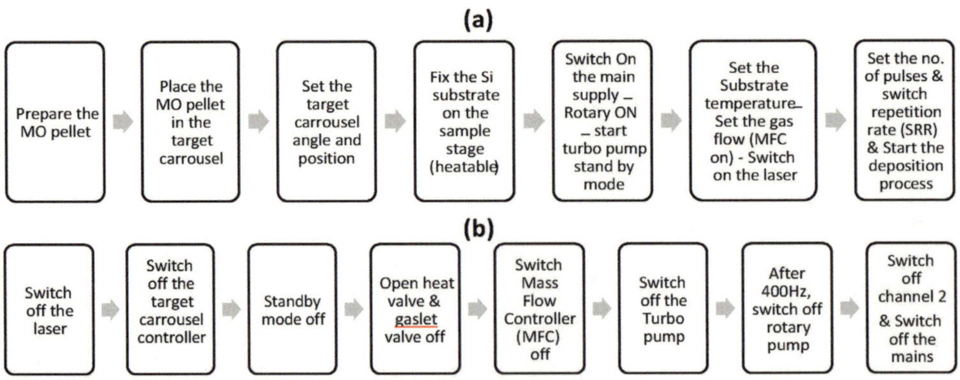

Figure 2.
Operation of PLD instrument (a) before and (b) after deposition.

iii. a precise control over the growth rate is enabled by the usage of laser beam and

iv. the transfer of the composition for most of the ablated materials show better congruence and stoichiometry.

3. Experimental details

Figure 3 shows the schematic of ZnO deposited on Si/glass substrate by pulsed laser deposition. In this work, pulsed laser deposition technique was used for growing high quality stoichiometric ZnO thin films. Laser from the source hits the ZnO target, a plasma plume consisting of molecules and atoms of zinc and oxygen start to deposit on the surface of the substrate. The plasma plume regime is known as the plasma perimeter. Mostly the plume will be centered in the course of the plasma perimeter. The plasma perimeter dependent upon the factors such as laser energy, laser repetition rate, temperature maintained, pressure and the distance between the target and the substrate. A detailed study of the effects of laser repetition rate, deposition and annealing temperatures on the electrical, optical and structural properties of the pulsed laser deposited ZnO thin films were carried out.

In the same way the preparation for other materials such as MoO_3, MoO_2, binary oxides ZnO/MoO_3, ZnO/TiO_2 and ZnO/V_2O_5 were carried out with pulsed laser deposition. Various analytical techniques such as SEM, UV–Visible spectroscopy, current–voltage characterization, ac complex impedance spectroscopy were also carried out to study the structural, morphological, optical and electrical properties.

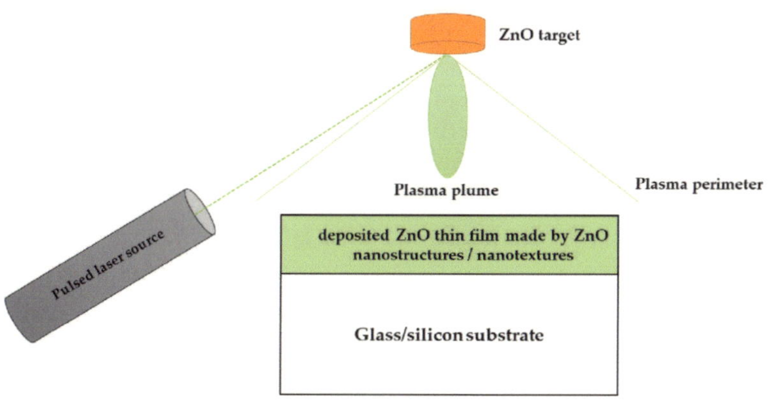

Figure 3.
Schematic of pulsed laser deposited ZnO thin film nanostructures/nanotextures on Glass/Si substrate [7].

4. Results and discussion

4.1 Investigation of ZnO nanostructures with the effects of laser repetition rate and deposition temperatures

The structure and properties of the transparent ZnO films deposited on glass substrates were then analyzed using X-ray Diffraction (XRD) as shown in **Figure 4** and confirmed the presence of hexagonal wurzite ZnO for the samples deposited at 300 °C, 450 °C and 600 °C [8].

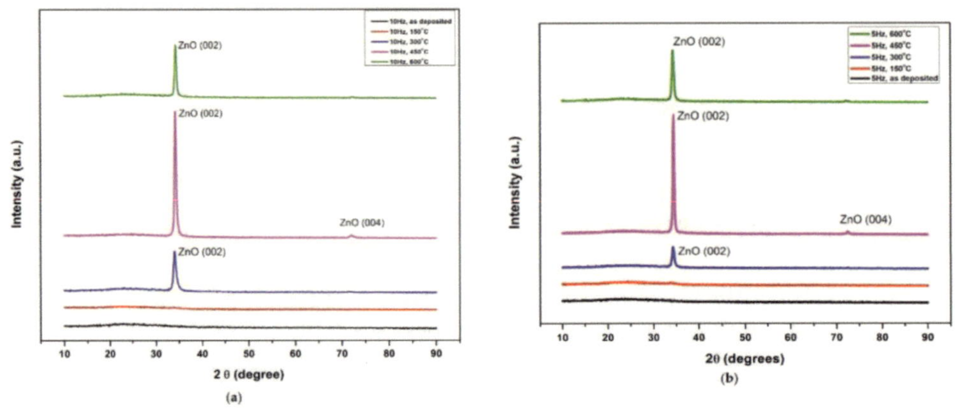

Figure 4.
XRD spectra of ZnO thin films (a) 10 Hz samples and (b) 5 Hz samples - (i) as deposited, (ii) at 150 °C, (iii) at 300 °C, (iv) at 450 °C and (v) at 600 °C.

It is also confirmed by the card no. 80–0074 of the Joint Committee on Powder Diffraction Standards (JCPDS) [9]. The XRD results provide the wurzite hexagonal crystal structure for the sample deposited at the temperature from 300-450 °C. The high intensity (002) plane preferentially grows in thin films above 300 °C. The (002) peaks become intense if the substrate temperature is increased from 300 to 600 °C. The grain size of the polycrystalline film increased with increasing the substrate temperature while deposition. **Table 2** provides the details of the bandgap energy calculated from the UV transmission spectra of the sample at various temperatures such as 25 °C (as deposited), 150 °C, 300 °C, 450 °C and 600 °C samples.

The transmission spectra for 10 Hz and 5 Hz samples are shown in **Figure 5**.

The ZnO thin films deposited at 10 Hz and 5 Hz shows an excellent transmittance and high transparency rate along with a decreasing energy bandgap as the temperature increases from as deposited sample at room temperature to 450 °C. The value of bandgap is estimated from fundamental absorption edge of the films. For the direct transitions, the absorption coefficient is expressed by

$$(\alpha h\nu)^2 = k\left(h\nu - E_g\right) \quad (1)$$

where k is constant, E_g is the energy bandgap, ν is the frequency of the incident radiation and h is Planck's constant. Because ZnO has a large exciton binding energy of 60 meV, an obvious exciton effect will always appear in the absorption spectra of high-quality ZnO films. As the ZnO thin film quality improves, a pronounced exciton absorption peak located at 3.1–3.3 eV was observed. The UV band was

Samples	UV emission center at 10 Hz (nm)	Bandgap energy (eV)	UV emission center at 5 Hz (nm)	Bandgap energy (eV)
As deposited	399 nm	3.209	400 nm	3.218
150 °C	395 nm	3.177	395 nm	3.177
300 °C	393 nm	3.161	394.8 nm	3.176
450 °C	393 nm	3.161	393.5 nm	3.165
600 °C	409 nm	3.290	394.3 nm	3.172

Table 2.
Bandgap energy of the samples deposited at various temperatures obtained from UV transmission spectra.

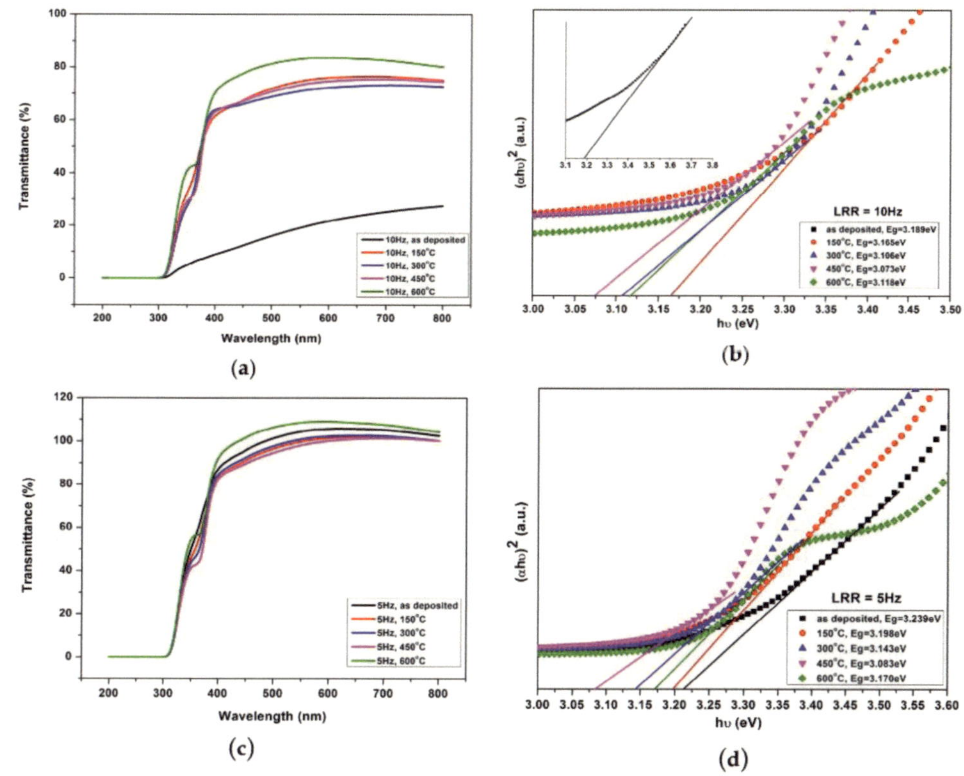

Figure 5.
UV-Transmittance (T%) spectra at various deposition temperatures (a) Wavelength Vs T% of 10 Hz samples, (b) Tauc-plot of 10 Hz samples (c) Wavelength Vs T% of 5 Hz samples and (d) Tauc-plot of 5 Hz samples.

assigned to be the free exciton. It is also observed that at 600 °C the energy bandgap started to increase slightly. The bandgap energies obtained from Tauc-plot of optical transmittance are tabulated in **Table 3**.

4.2 Investigation of MoO_3 and MoO_2 nanostructures

Figure 6 shows the FESEM images of the pulsed laser deposited with various thin film of molybdenum oxide. The sample deposited at 450 °C and using the O_2 gas atmosphere, near the edges and unfilled region, hexagonal (100) nanotubes, along with neatly arranged orthorhombic (040) molybdenum trioxide (MoO_3) structures is shown in **Figure 6a**. Followed with **Figure 6b** shows the sample deposited at 450 °C using the O_2 gas atmosphere, uniformly arranged orthorhombic (040) structures of the MoO_3 seen. Then, **Figure 6c** shows the sample stored at

Samples	E_g at 10 Hz (eV)	E_g at 5 Hz (eV)
As deposited	3.189	3.239
150 °C	3.165	3.198
300 °C	3.106	3.143
450 °C	3.073	3.083
600 °C	3.118	3.170

Table 3.
Bandgap energy obtained from Tauc plot of optical transmittance.

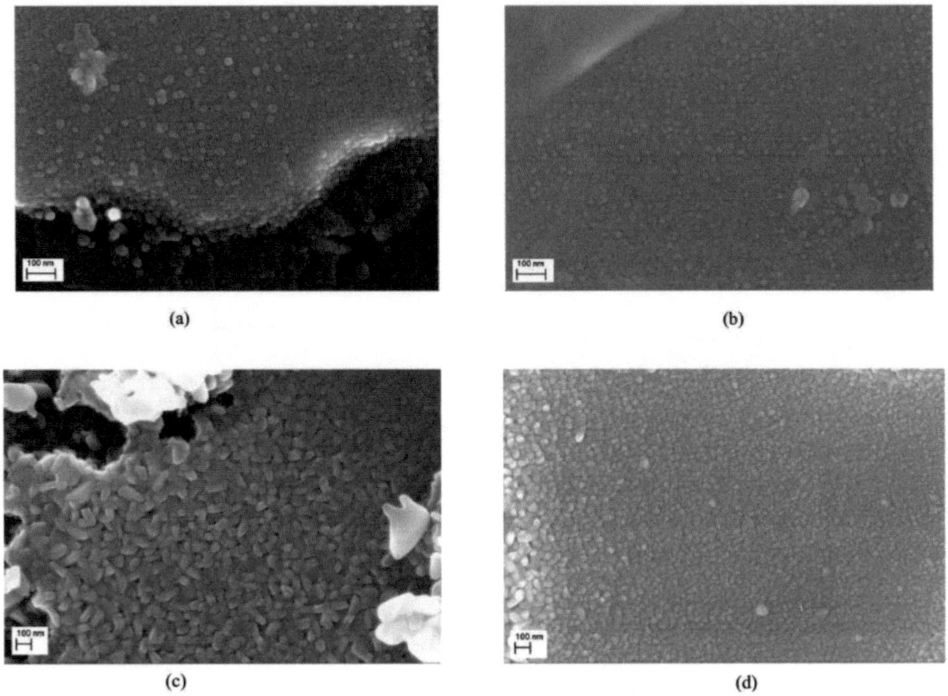

Figure 6.
Field Emission Scanning Electron Microscopy (FESEM) images of molybdenum oxides thin films deposited by PLD. (a) and (b) MoO_3 at 450 °C (c) and (d) MoO_2 at 450 °C.

450 °C under the argon gas atmosphere, near the unfilled region, neatly arranged monoclinic molybdenum dioxide (MoO_2) structures. Finally, the sample deposited at 450 °C using the argon gas atmosphere, which shows uniformly arranged monoclinic MoO_2 structures is shown in **Figure 6d** [10].

Figure 7 shows the XRD spectra of the pulsed laser deposited molybdenum oxide thin films at the various conditions. The sample deposited at the room temperature was the as-deposited thin film which exhibits the amorphous nature. Then the sample was deposited at a substrate temperature of 450 °C and 600 °C, which exhibit orthorhombic structure with 020 and 040 peaks and are confirmed with the ICSD 80577. Then the deposition duration was increased from 2 minutes to 20 minutes, with the O_2 atmosphere, which results in orthorhombic structures and hexagonal structures and the peaks are found at 100 and 211 respectively. Then the next sample when deposited with same temperature and duration under the Ar atmosphere, the XRD spectra shows the monoclinic structures and confirms as in **Figure 4** the presence of MoO_2 with 110, 020 and 220 peaks with the ICSD 23722 and JCPDS 65–5787. When the sample deposition temperature increased from 450 °C to 600 °C, the structure crystallization takes place [11].

4.3 Investigation of conduction mechanism by ac complex impedance spectroscopy for PLD binary oxides ZnO/MoO_3 thin films

The complex impedance (Z" vs. Z') plots of pulsed laser deposited binary oxides ZnO/MoO_3 (ZMO) thin films are displayed in **Figure 8**. The response of a measurement in a complex impedance plot enables us to separate two contributions which appear in the form of semicircles arcs. We know that, for a semiconducting material having interfacial boundary layers (grain-boundary) and the figure exhibits

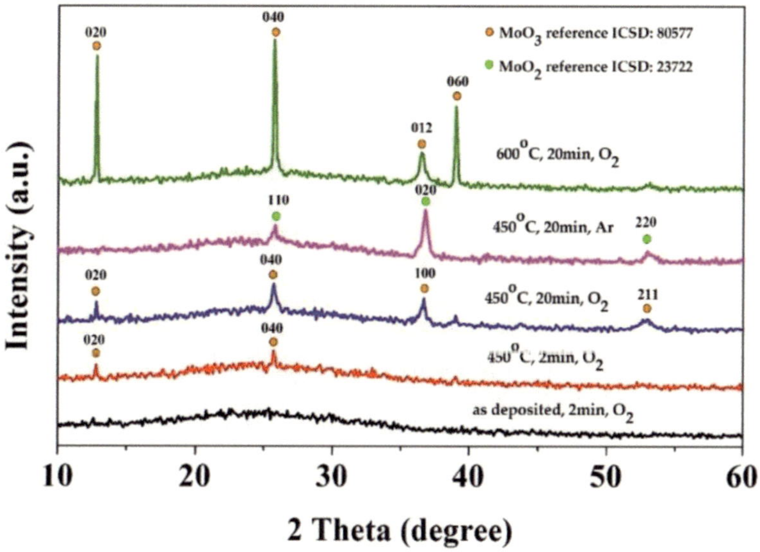

Figure 7.
XRD spectra of pulsed laser deposited Orthorhombic MoO_3 and monoclinic MoO_2 thin films prepared at various deposition conditions.

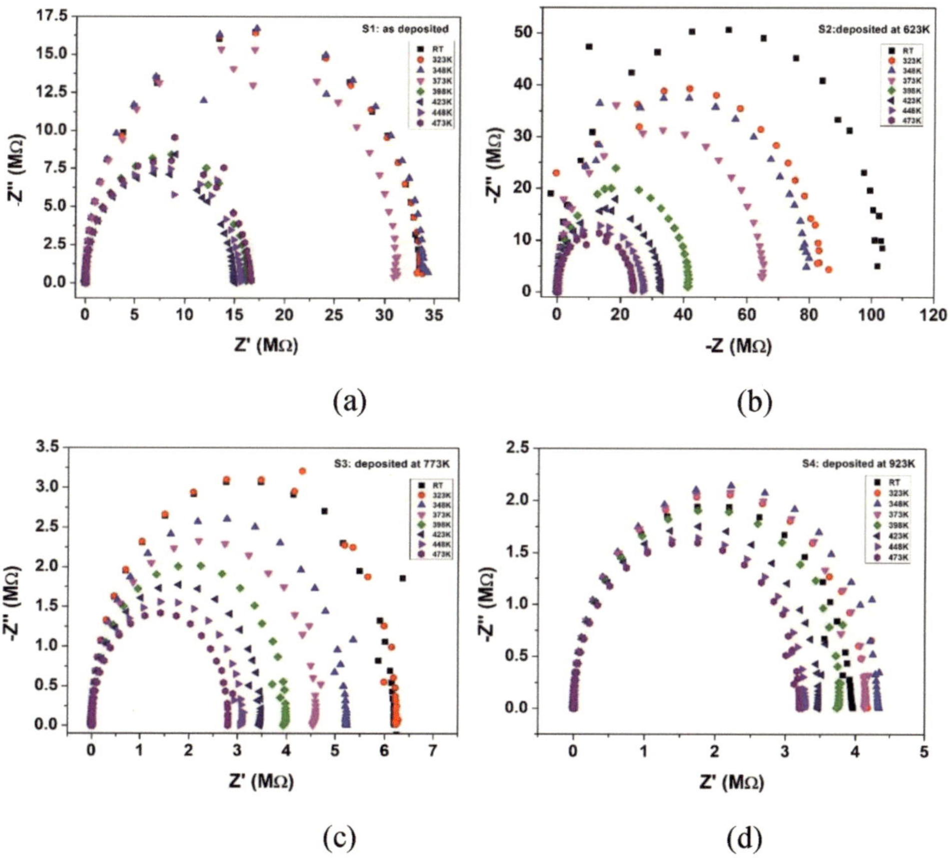

Figure 8.
The complex impedance (Z" vs. Z') plots of ZMO thin films (a) 298 K (b) 623 K (c) 773 K and (d) 923 K.

semicircles that are deformed and depressed with their centres below the real axis as the temperature range investigated, complex curve consists only one depressed semi-circle and its centre lies below the real axis. Furthermore, depressed arc is typical for a dipolar system involving distribution of relaxation time. Moreover, it is noted that the diameter of semi-circles decreases with increase in temperature which also refers to the decrease in the resistivity. The equivalent circuit for this sample is a series Resistor-Capacitor (RC) circuit. Whereas the semi-circled impedance samples possess the characteristics which are attributed to the semiconductor behaviour, in which the electrical conduction process is thermally activated.

The equivalent circuit for these depressed semi-circles of the ZMO thin films may be described by a parallel connection of an ohmic resistor R and a capacitor C, also known as Randles circuit. Here the capacitor is replaced with a constant phase element (CPE) associated with both the resistors and capacitors [12]. Indeed, in the present study, the complex plane plot can be described by the Nyquist plot, which is given by

$$Z = \frac{R}{\left[1 + (j\omega\tau)^{\alpha}\right]} \tag{2}$$

where ω is angular frequency, τ = RC is the relaxation time and α is a parameter that characterizes the distribution of relaxation times with values ranging from 0 to 1. When α is zero, the relaxation is said to be Debye relaxation and when α is greater than zero, the relaxation times are said to have distribution.

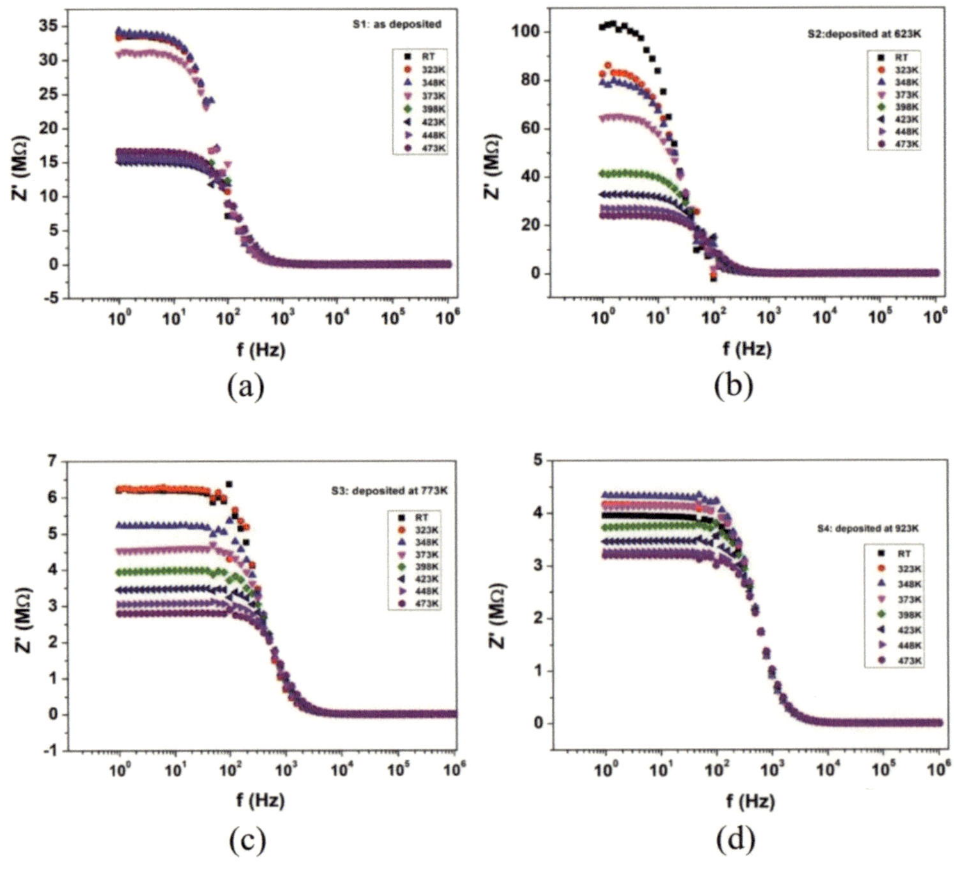

Figure 9.
Frequency dependence of the real part (Z') of the complex impedance. (a) As deposited ZMO, (b) ZMO deposited at 623 K, (c) ZMO deposited at 773 K and (d) ZMO deposited at 923 K.

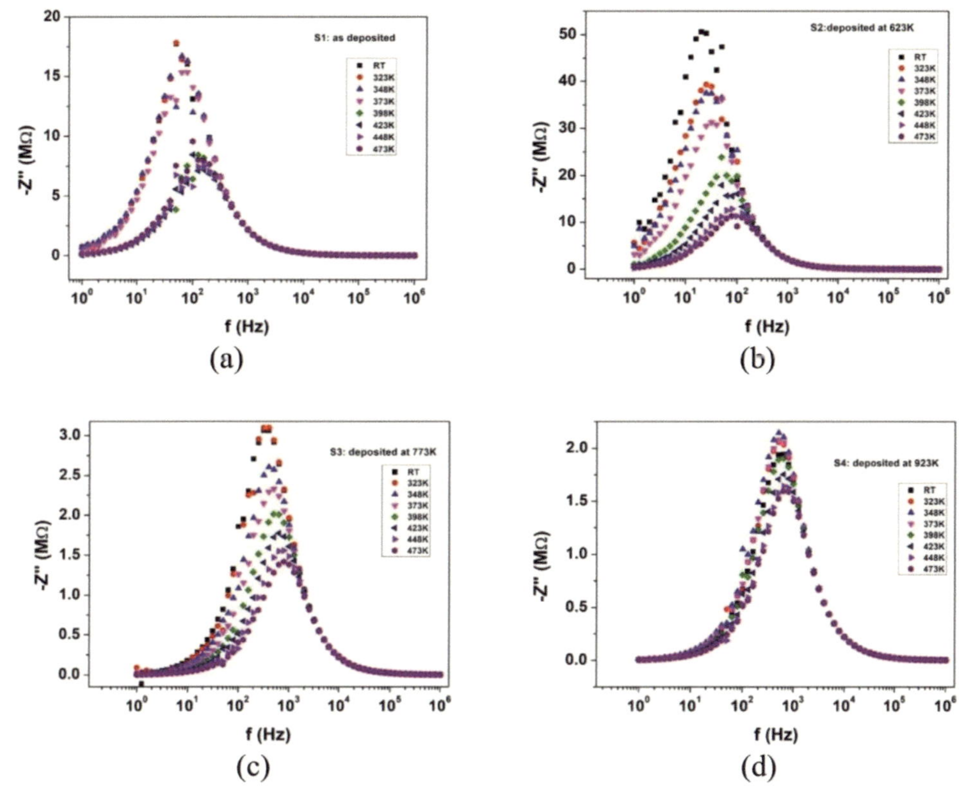

Figure 10.
Imaginary part of impedance (Z") as a function of frequency. (a) As deposited ZMO, (b) ZMO deposited at 623 K, (c) ZMO deposited at 773 K and (d) ZMO deposited at 923 K.

Figure 9 shows the frequency dependence of the real part (Z') of the complex impedance Z* = Z'-jZ" for temperatures ranging from 298 K to 923 K [4].

It is observed that as there is a rise in the temperature and the frequency, the real impedance magnitude Z' decreases and hence the materials show a negative temperature coefficient of resistance (NTCR) and are attributed to a semiconductor behaviour. It is also noticed that the value of Z' for all the temperatures merges towards the high frequency due to the space charge dependent behaviour. The charge carriers are settled on the portion of the grain boundaries with sufficient energy to overcome the barrier with the increase in the temperature and hence we can emphasize as in conductivity. It can be seen that the peak position shifts from lower to higher frequencies with increasing temperatures. But the maximum value of the imaginary part Z''_{max} values decreases due to thermally activated dielectric relaxation process as shown in **Figure 10** [4].

The same investigation carried out for the binary oxides ZnO/V_2O_5 and ZnO/TiO_2 and its performance as dielectric material with respect to wide frequency range and wide temperature range. The behaviour is compared and studied with respect to various deposition temperatures [13].

5. Conclusions

The nanostructured thin films metal oxides such as ZnO, MoO_3, MoO_2, binary oxides ZnO/MoO_3, ZnO/TiO_2 and ZnO/V_2O_5 were prepared using pulsed laser deposition technique. The ZnO thin film nanostructures are seen through the SEM/FESEM at the different deposition temperature. Through the increase of deposition temperature to 450 °C, the orthorhombic MoO_3 and monoclinic MoO_2 are prepared at the O_2 and Ar gas deposition atmosphere and confirmed using the FESEM and XRD. The binary oxides ZnO/MoO_3, ZnO/TiO_2 and ZnO/V_2O_5 shows amorphous nature even at high deposition temperature. The thin films dielectric properties, electric modulus and impedance properties were analysed using ac complex impedance spectroscopy. The dielectric relaxation process is thermally activated for all the samples suitable for channel applications in FET devices. They confirm them to possess the nature of semiconducting property by the deformed semi-circle for the impedance Nyquist plots.

Author details

Cyril Robinson Azariah John Chelliah[1*] and Rajesh Swaminathan[2]

1 Department of Nanotechnology, Institute of Electronics and Communication Engineering, Saveetha School of Engineering, Saveetha Institute of Medical and Technical Sciences, Saveetha University, Chennai, Tamil Nadu, India

2 Department of Physics, School of Sciences, Arts, Media and Management, Karunya Institute of Technology and Sciences, Coimbatore, Tamil Nadu, India

*Address all correspondence to: cyrilrobinsonazariahj.sse@saveetha.com

References

[1] Rembert, T., Battaglia, C., Anders, A., & Javey, A. (2015)"Room Temperature Oxide Deposition Approach to Fully Transparent, All-Oxide Thin-Film Transistors", Advanced Materials, 27, 6090-6095.

[2] Babu, S. S., Moni, D. J., Padickala, P. J., Azariah, J. C. R., & Rajesh, S. (2018, March). Fabrication and Characterization of PLD Deposited Crystalline Zno as Channel and Amorphous Zno as Gate Dielectric of the Thin Film FET. In *2018 4th International Conference on Devices, Circuits and Systems (ICDCS)* (pp. 301-304). IEEE.

[3] Challam, A., Chelliah, C. R. A. J., Nirmal, D., & Swaminathan, R. (2018). Design and Fabrication of GaAs Based MOSFET by Physical Vapor Deposition Method. Materials Focus, *7*(4), 453-458.

[4] Chelliah, C. R. A. J., & Swaminathan, R. (2020). Enhanced Structural, Optical and Electrochemical Properties of Pulsed Laser Deposited Binary Zinc and Molybdenum Oxide Nanostructured Thin Films. In *Key Engineering Materials* (Vol. 853, pp. 68-72). Trans Tech Publications Ltd.

[5] Pandiyarajan T., Mangalaraja R.V., Karthikeyan B., Mansilla H.D., Gracia-Pinilla M.A. (2017) Spectroscopic Investigation on rGO:ZnO Composites Nanostructures. In: Ebenezar J. (eds) Recent Trends in Materials Science and Applications. Springer Proceedings in Physics, vol 189. Springer, Cham. https://doi.org/10.1007/978-3-319-44890-9_7

[6] Zhang, Yufei, Laiquan Li, Haiquan Su, Wei Huang, and Xiaochen Dong. "Binary metal oxide: advanced energy storage materials in supercapacitors." Journal of Materials Chemistry A 3, no. 1 (2015): 43-59.

[7] Chelliah, C. R. J., & Swaminathan, R. (2018). Pulsed laser deposited hexagonal wurzite ZnO thin-film nanostructures/nanotextures for nanophotonics applications. Journal of Nanophotonics, *12*(1), 016013.

[8] Chelliah, C. R. A. J., & Swaminathan, R. (2018). Study of the pulsed laser deposited ZnO thin films and its electrical performance as n-Channel in MOSFET. Journal of Nanoelectronics and Optoelectronics, *13*(5), 708-714.

[9] Kim, S. H., Badran, R. I., & Umar, A. (2017)"Fabrication of ZnO Nanorods Based p – n Heterojunction Diodes and Their Electrical Behavior with Temperature", Journal of Nanoelectronics and Optoelectronics, 12, 731-735.

[10] Chelliah, C. R. A. J., Ponmudi Selvan T., Samuel Rajasekar M., Sheebha, I., Vidhya, B., & Rajesh, S. (2018, March). Pulsed Laser Deposited Molybdenum Oxides (MoO 3 & MoO 2) Thin Films for Nanoelectronics Device Application. In *2018 4th International Conference on Devices, Circuits and Systems (ICDCS)* (pp. 42-47). IEEE.

[11] Balaji, M., Chandrasekaran, J., Raja, M., & Rajesh, S. (2016)"Structural, optical and electrical properties of Ru doped MoO3 thin films and its P–N diode application by JNS pyrolysis technique", Journal of Materials Science: Materials in Electronics, 27, 11646-11658.

[12] Ait Ahsaine H., Zbair M., Ezahri M., Benlhachemi A., Arab M., Bakiz B., Gavarri J.-R. (2015) "Rietveld refinements, impedance spectroscopy and phase transition of the polycrystalline ZnMoO4 ceramics", Ceramics International, 41, 15193-15201.

[13] Chelliah, C. R. A. J., & Swaminathan, R. (2019). Improved optical absorption, enhanced morphological and electrochemical properties of pulsed laser deposited binary zinc and vanadium oxide thin films. *Journal of Materials Science: Materials in Electronics*, 1-11.

Section 2

Laser Surface Modification, Machining and Nanoparticle Formation

Chapter 4

Laser Surface Modification of Materials

Natarajan Jeyaprakash, Che-Hua Yang and Durairaj Raj Kumar

Abstract

The metallic materials such as steel, iron, titanium and nickel alloys etc., are extensively used in the automobile, marine, biomedical, aerospace, chemical industry and power generation sector. However, the poor surface properties restricted their wide usage in many applications. Therefore, the surface properties need to be enhanced through novel treatments without affecting the bulk. In recent years, laser surface modification attracts more due to their inherent properties. The laser based surface altering process is appropriate to modify the metallic surfaces in terms of their flexibility, simple operation and process economy. Laser surface modification includes; surface hardening, melting, alloying, cladding and texturing. Thus, from a process engineering, metallurgical reasons and tribologist view point, the laser surface modification process can be recognized as an important topic.

Keywords: laser hardening, melting, alloying, cladding and texturing

1. Introduction

In the 20th century, Laser surface alteration played a major role in enhancing the material surface properties. Among the number of ways to enhance the material properties, laser based surface alterations are used to enhance a better physical property in the machined surface and improved the component performance. The high power Neodymium Yttrium-Aluminum-Garnet (Nd: YAG) laser, carbon-di-oxide (CO_2) laser and excimer lasers are used to perform the laser surface treatment which is expensive, popular and operate at pulsed mode or continuous wave mode. These lasers are used to heat the near-surface area of the finished components for enhancing the properties. The laser surface modifications have the ability to control the amount of heat energy to work material with high directionality. The purpose of a surface hardening by laser is to improve the component wear properties. The laser surface hardening is defined as the heat energy from the laser beam that directly heated the component surface at a very short interval period without melting the work material. The heat input to the component surface is the reason for creating the tough and fine-grained structure in the hardened surface. The risk of crack forming is very low due to the self-quenching process. The laser surface melting (LSM) is heated to its melting point through a high power laser beam and rapidly solidified. The aim of LSM is to refine the surface microstructure, homogenization of composition, dissolution of precipitates. The LSM is also used to improve the corrosion resistance of steel and iron. The minimization of intergranular corrosion is possible through LSM by avoiding the carbides formation during subsequent homogenization and sensitizing treatment. The laser surface alloying (LSA) is defined as the high heat energy used to melt the metal coating through laser and a portion of underlying substrate. This technique is used to form highly resistant gradient layers on the metal surface. The major benefit of this technique is sudden heating followed by cooling and the surface properties are improved. The laser cladding (LC) is a coating method that the surface melting and new material layer formation by addition of material are simultaneously processed in the substrate at the same time by using the laser power. The desired surface properties are achieved after solidification. The large component surface properties are easily increased by using LC. The complete metallurgical bond is necessary between the melting of substrate and forming of a new material layer at the interface. The laser surface texturing (LST) is defined as the process in which the change of material surface properties by modifying its texture and roughness. The laser beam is used to create the micro patterns on the surface by laser ablation. The micro patterns are created on the surface in various shapes such as dimples, grooves and free forms with precise dimension. This process is mostly used in biomedical applications.

The different types of laser have different abilities to perform the process on materials. All the lasers are producing the heat energy and the laser beam wavelength is majorly affecting the performance of materials. Generally, the total laser heat energy is supplied to work material in which can be divided into two ways such as the fraction of heat energy is observed by work material and remaining heat energy is reflected to the environment. This happens during the surface hardening by laser. The supply of heat energy to polished metal surface components is depending upon the heat absorbability of work material and wavelength of irradiation. Generally, the short wavelength has higher absorptivity. Hence, the Nd: YAG laser (λ = 1.064 μm) has produced the higher absorbing ability beam to work material than the CO_2 laser (λ = 10.6 μm) for surface hardening of steel. In order to increase the CO_2 laser absorbility (high wavelength) to work material, the coating or painting is required in the work material prior to the CO_2 laser surface hardening. Therefore, the Nd: YAG laser surface hardening better than CO_2 laser surface hardening because the Nd: YAG laser has short wavelength and produces a high absorbing rate to work material. The Nd: YAG laser produces heat energy to work material which is transferred through fiber cable whereas CO_2 laser is impossible. The inert gases, helium, neon and argon are used to eliminate the atmospheric contamination. In order to reduce the wavelength of a laser, an excimer laser is developed with very short wavelength. This laser can be used to micromachining on medical parts. In this chapter, laser surface hardening, laser surface melting, laser surface alloying, laser surface cladding and laser surface texturing have been discussed to improve the microstructure, hardness and wear resistance of mechanical components.

2. Laser surface hardening (LSH)

The laser surface hardening is defined as the heat energy from the laser beam which is directly impacted to the finished component surface for improving the wear resistance. The component life is increases without affecting the bulk material. During the hardening process, the surface layer is heated up to hardening temperature under the short period of time. The quenching is a necessary process to achieve the hard martensite phase in the heated surface. Thereby, the component surfaces are hardened by laser and achieve the high wear resistant surface with desired bulk properties. The components such as gear teeth, gears, shafts, camshafts, axles, cylinder liners, valve guides and exhaust valves showed with higher stresses due to laser surface hardening. The type of work materials, cast iron, die steel and medium-carbon steel are also required the laser surface hardening for better performance. The mass-production industries, automobile components and electronic parts are performed the laser hardening on the component surfaces [1]. The desired component performances are mainly depending upon the selection of laser process parameters such as power, scanning speed, pressure, beam shape and material properties. Now-a-days, in order to improve the surface quality of components, the number of surface treatment are commercially available to obtain the unique material properties. For example, the I-section rail (railway) is fabricated by hot rolled processes which have non-uniform properties in the flange and web. The I-section beam is shown in **Figure 1**. The flanges have been designed to withstand high stress whereas the web designed to withstand the least stress. The flange thickness is greater than the web thickness and stress developed in the I-section is within the allowable limit. The point is the different cross section of flange and web has produced the non-uniform properties. Hence, the laser surface hardening is required for achieving the uniform properties over the flange and web.

The laser surface transformation hardening process is performed to obtain the required depth and width for steel material. The accurate parts are made of medium carbon steels which require the laser surface hardening. The small and complex components are easily surface hardened by laser. This is because of the high rate of cooling effects to increase the hardness rate in the quenching process [2]. Therefore, LSH is a better process compared to flame and induction hardening processes. The quenching process is suddenly reducing the work material temperature by using water, oil or air to get certain material properties through the phase transformation. Therefore, a comparative study is made between the laser quenching and conventional quenching on steel to study the hardness and wear rate. The conventional quenching and tempering is carried out by using the temperature of 1198 K for 4.5 h and temperature of 523 K for 4 h respectively. The air, 10 kW CW diode laser, 3.5 mm spot diameter and 168 mm/s linear speed are used in the laser treatment. The laser quenched and conventional quenched sample for 25 μm distance from the surface, the produced hardness is 600 $HV_{0.1}$ and 625 $HV_{0.1}$ respectively. The laser quenched sample has 0.4 mm^3/N-m wear rate which is lesser than the conventional quenched sample of 0.6 mm^3/N-m wear rate at 500 m sliding distance [3]. The wear and microhardness studies are performed on 40CrNiMoA steel by using laser quenching and high-frequency quenching.

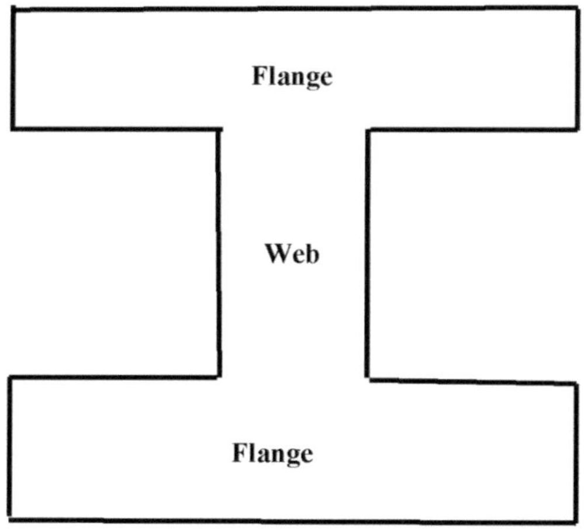

Figure 1.
Schematic of I-section beam used in rail.

A 2 kW CW CO_2 laser, 1400 W laser power, 35 mm/s traverse speed, 60 degree incident angle, black organic absorbent coating, 0.9 m^3/h gas flow rate and 10 mm defocusing distance are used in the laser treatment. The hardness of the quenched groove surface reached 750 HV and is substantially higher than that resulting from the high-frequency quenching method. The results of wear testing showed that the wear resistance of laser quenched specimens is 1.3 times higher than that of a high-frequency quenching specimen [4]. Comparisons were made between the gray cast iron (GCI), laser hardened quench-tempered GCI and conventional austempered GCI specimens based on the hardness and wear loss. The air, CW Nd: YAG laser, 2 mm laser spot, 22 mm defocused distance, 2 mm/s scanning speed, 6 Hz frequency, 120 A current and 8 ms pulse duration is used for laser hardening. The hardness of the laser hardened zone with ledeburitic structure is approximately 68 HRC. The quenching-tempered GCI specimen showed higher wear resistance than untreated GCI specimen [5].

The advantage of laser surface hardening is listed below

- The lower level of heat energy is used to work material compared to conventional surface heat treatment.
- The input laser energy is controlled by varying the process parameters such as power, scanning speed, defocus, different shapes of lenses and mirrors.
- The hardened surface is obtained through self-quenching of the heated surface layer.
- The work material is made under the hardening and quenching process resulting in cleaning of work material is not required.
- The beam guidance is automatically controlled over the work material.
- The surface heat treatment is specifically performed on small parts and complex parts.

The disadvantage of laser surface hardening is listed below

- High initial capital cost
- Skilled operators are needed
- Surface preparations are required in difficult areas.
- Radiation protection is required
- Material hardness and wear

The performance of the components such as hardness and wear resistance of work materials are mainly focused in the laser surface hardening. This is depending upon the material type, material properties, and types of processing on materials. The desired properties of work materials are obtained through proper selection of laser surface treatment and optimization. In order to improve the durability of mechanical components namely gears, engine valve, brake drums and camshaft are highly needed the LSH. The induction hardening is one of the surface hardening process which is shown in **Figure 2**. It is performed to achieve the uniform

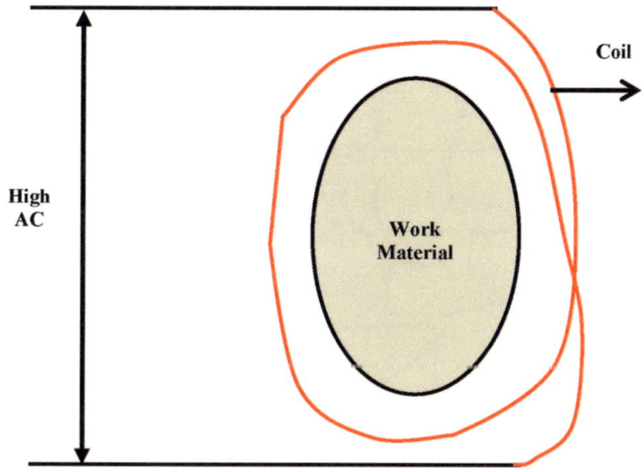

Figure 2.
Schematic of induction hardening.

microstructure and good wear resistance which is higher implementation cost compared to laser surface hardening.

In this induction hardening, the depth of hardening is mainly depending upon the resistivity (ρ), frequency (v) and magnetic permeability (μ). The work material is placed inside the coil and supplies the high frequency. The surface is hardened by skin effect.

$$d = \sqrt{\frac{\rho}{\mu v}} \tag{1}$$

The laser surface hardening can be performed on the components either partially or fully depending upon the application of the components. Specifically, the load bearing component is subjected to high surface wear. Hence, the laser surface hardening is required on the load bearing component surface. Therefore, the load bearing component is hardened by laser, the surface has produced a high hardenability and fine microstructure [6]. The service life of crankshaft and camshaft are made on EN18 steel in which properties are improved by diode laser surface hardening with beam diameter of 3 mm, velocity of 1 m/min and power of 1.5 kW. The argon gas is used as shield gas [7]. The advantages of induction hardening are localized areas heat treated, minimal surface decarburization, surface oxidation, slight deformation, improved fatigue strength and low operating cost. The disadvantages of induction hardening are high capital investment. The advantages of laser hardening are described as non-hardenable steels are surface hardened, higher hardness obtained than conventional hardening, eliminating dimensional distortion, no protective atmosphere required and very long and irregular shapes easily hardened. The disadvantages of laser hardening are high initial and working cost and difficult to harden the high alloy steel. The schematic diagram of substrate and laser processed materials are shown in **Figure 3(a)** and **(b)**. The parent substrate has coarse and uneven equiaxed grains. The laser processed work material showed the hardened depth varying from top surface to 200 μm depth. The depth of hardening increases with grain size increases from finer to coarser. The curved surface is formed at top surface due to the low scanning speed produces more evaporation in the laser melted surface. The laser process parameters, power of 1.5 kW, beam diameter of 3 mm, scan speed of 1 m/min and interaction time of 0.18 s are used to obtain the desired hardness. The Nd: YAG laser and argon gas with flow rate of 20 L/min is used in the laser surface hardening. The hardness decreases from 955 HV to

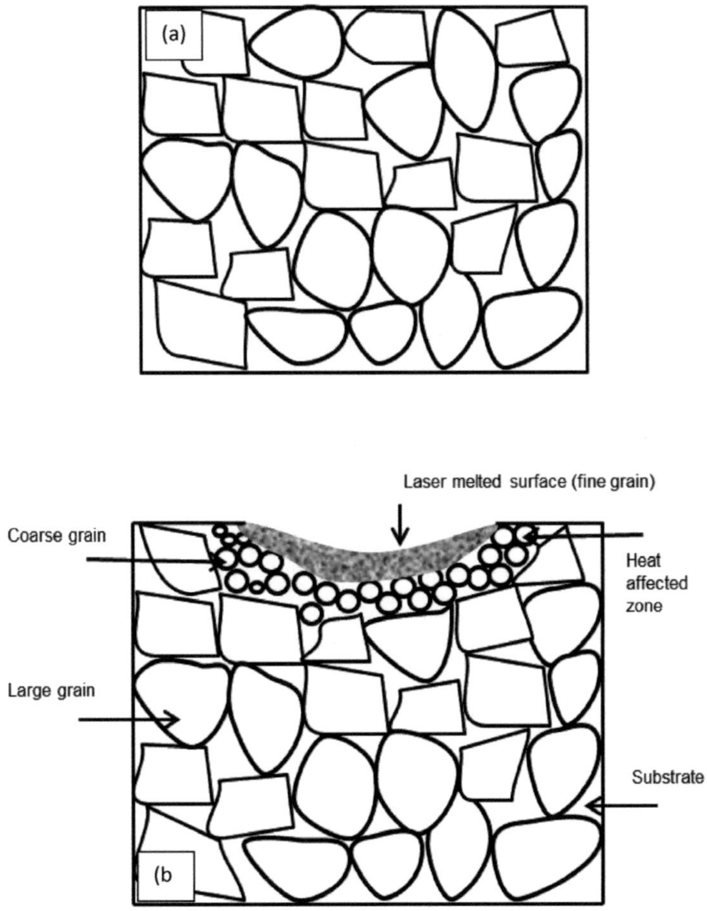

Figure 3.
(a) Schematic of; (a) as received tool steel microstructure, (b) laser surface hardened tool steel with modified structure.

236 HV which is obtained by varying distance from top to 200-micron depth and it is shown in **Figure 4**. This is due to the grain size refinement [8]. The 5 kW CW CO_2 laser, power ranging from 1.1–2.5 kW, traverse speed ranging from 6 to 15 mm/s and spot size of 6.3 mm, 2.27 mm, 4.63 mm and 1.2 mm are used to harden the various carbon steel. The argon gas is used as shielding gas. The traverse speed has mostly affecting the hardness. The carbon percentage increases, the average hardness value also increases. The C-45 steel has produced higher hardness. The hardness of the material was improved by minimizing the diameter of spot size [9]. Further, conventional type laser surface treatment is performed on large surface areas and irregular hardness was observed on the machined component. In order to overcome irregular hardness, a laser overlapping method is used in the laser transformation hardening which is presented in **Figure 5**.

After the laser treatment, the laser hardened zones are divided into three sections such as hardened zone, transition zone and heat affected zone which is shown in **Figure 6**. A study on the effect of process parameters on surface hardness splined shafts is performed by using laser surface hardening. The fiber laser, power varying from 1900 to 2500 W, scanning speed varying from 2 to 6 mm/s, rotation speed varying from 1500 to 2500 rpm, the flank tilt angle of spline tooth varying from 15 to 20 and tooth depth of spline shaft varying from 2.5–3.5 are used in the laser

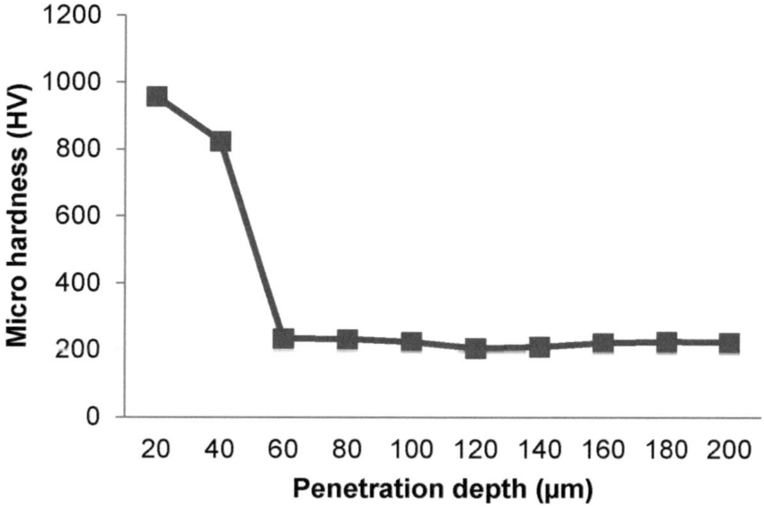

Figure 4.
Microhardness variation from top surface to substrate through LSH.

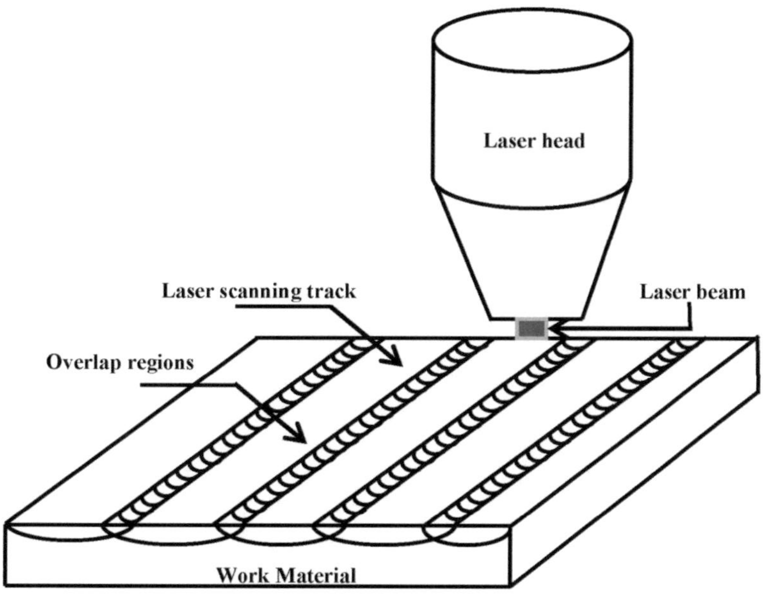

Figure 5.
Schematic of laser transformation hardening.

hardening of spline shaft. The result found that the maximum hardness is observed by using the power of 2500 W, scanning speed of 2 mm/s, rotational speed of 2500 rpm, the flank tilt angle of spline tooth of 20° and tooth depth of spline shaft of 3.5 mm [10]. An investigation on the underwater hardening of AISI 1055 steel is carried out using lasers. A 250 W CW Ytterbium based fiber laser, focal length of 300 mm, defocus distance of 10 mm and traverse speed varying from 1 to 100 mm/s are used in the laser surface hardening. The result found that the higher surface roughness is obtained in the underwater welding compared to conventional laser hardening due to the additional cooling effect in the underwater [11].

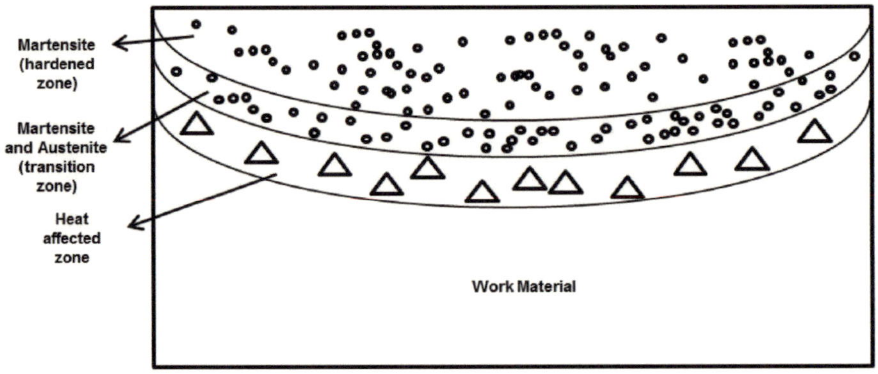

Figure 6.
Schematic of different zones of laser transformation hardening.

3. Laser surface melting (LSM)

Laser surface melting is one of the surface alteration processes that the surface of the substrate is melted and rapidly solidified to form the fine microstructure and improving the mechanical properties without changing the bulk properties and without addition of any metallic elements. The piston, valve and sliding parts are made of magnesium alloys, which are used in the automobile components and energy saving material. The application and limitation of magnesium alloy is decided by properties. In order to improve the tribological and mechanical properties, the laser surface melting process is focused on magnesium alloy. In the conventional heat treatment of HSS materials are presented the retained austenite, which transforms into brittle martensite during service. But, the life of high-speed tool steel is increased by using LSM. The schematic view of LSM is shown in **Figure 7**. The LSM treatment are carried out using a 2 kW fiber laser with 1.06 μm wavelength, laser power of 1500 W, the laser scanning speed of 600 mm/min and the distance between the laser head, spot size of 3 mm, shielding gas pressure of 0.3 MPa and the specimen

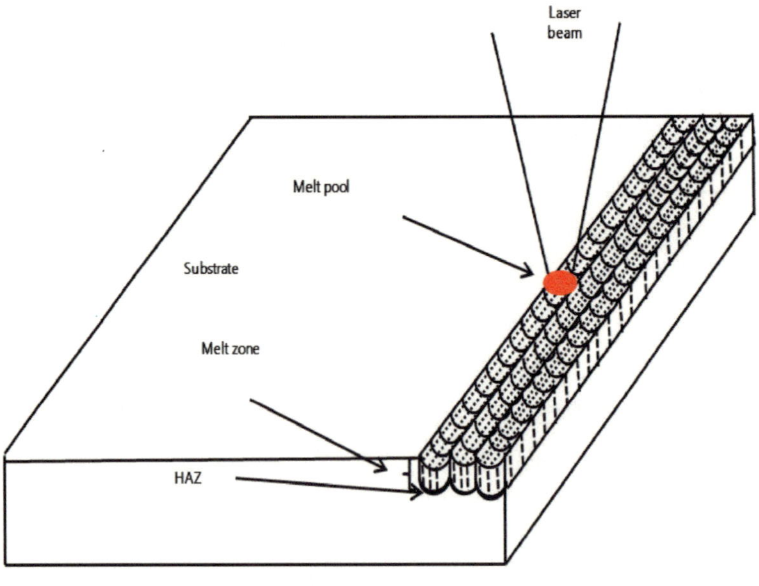

Figure 7.
Schematic view of laser surface melting.

surface of 12 cm are used in the LSM. The microhardness and corrosion resistance of magnesium alloy is also improved by using LSM with electromagnetic stirring [12]. In order to enhance the microhardness of melted substrate, the LSM process parameters effect of hardness of magnesium alloy is studied. The CW CO_2 laser, beam diameter of 4 mm, argon gas of 6 l/min, speed varying from 100 to 400 mm/min and power varying from 1.5–3.0 kW are used in the process. The result showed that the melt depth of magnesium alloy is directly proportional to the laser power and inversely proportional to the scan speed. Laser surface melting enhances the microhardness of the melted zone by 2–3 times than the substrate [13]. The laser processed hardness of high speed tool steel and magnesium alloy is decreased from as-received substrate by increasing distance from the melting surface which is shown in **Figure 8**. This is due to the refined, solid solution strengthening and uniform microstructure. The LSM is also performed in electric contact material of Cu-50Cr. The 1 kW CW Nd: YAG laser, power density varying from 10^6 to 10^7 W/cm^2, scanning speed of 6000–10,000 mm/min and argon gas are used in this process. From the analysis found that the microhardness and withstanding voltage of Cu-50Cr are significantly improved by using LSM [14]. The effects of LSM process parameters are affecting the microstructure and hardness of AZ31B magnesium alloy substrate. The result found that the grain size in the fused layer increases by increasing power. The schematic diagram of as-received magnesium alloy is shown in **Figure 9a**. The effects of different power on microstructure of layer fused layers are shown in **Figure 10b–e**. The Nd: YAG laser power varying from 1600 to 2200 W, laser beam scanning velocity of 900 mm/min, laser beam spot diameter of 4 mm, number of superimposed tracks of 9, overlap ratio of 15%, and argon flow rate of 25 mL/min are used in the process. The depth of the metal pool and grain size is increased by increasing the power. This is due to the grain growing freely in the higher metal pool depth compared to smaller metal pool depth. The reason for increasing the hardness and wear resistance are

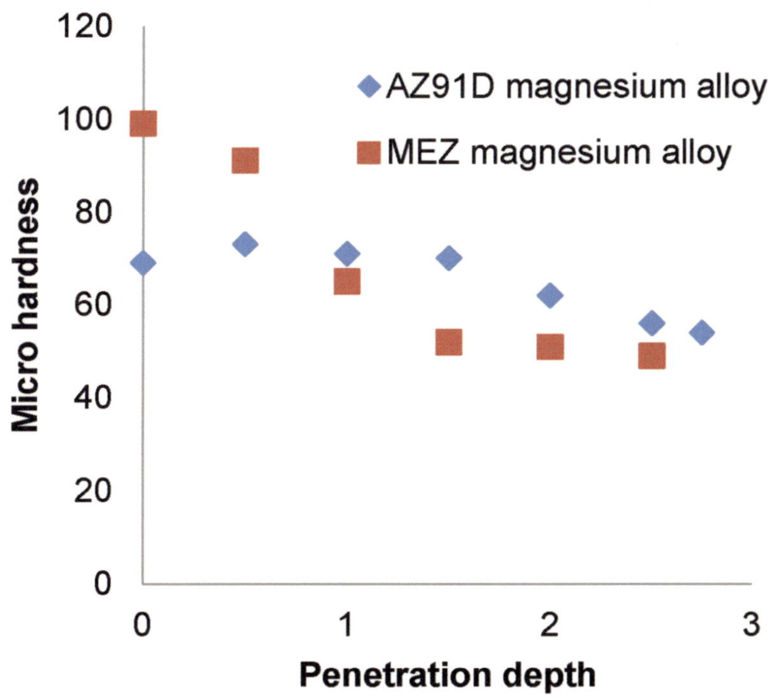

Figure 8.
Microhardness variation of magnesium alloy.

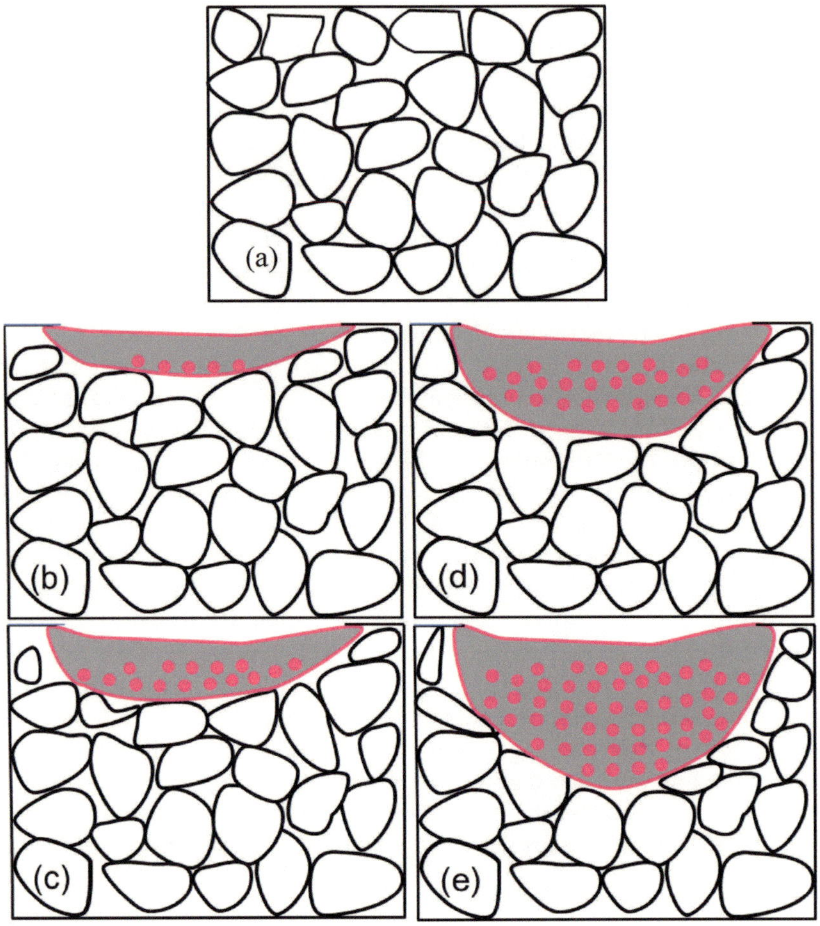

Figure 9.
Schematic diagram of (a) As received AZ31B magnesium alloy, microstructure of laser fused layer of (b) laser melted at 1600 W, (c) laser melted at 1800 W, (d) laser melted at 2000 W.

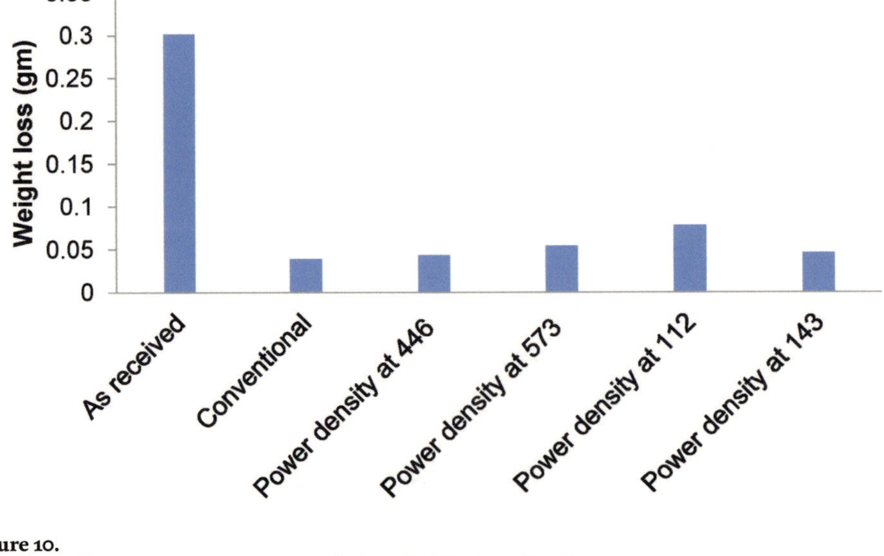

Figure 10.
The effect different heat treatment on weight loss of AISI M2 tool steel.

due to the grain refinement, high dislocation density and dispersive distribution of β- $Mg_{17}Al_{12}$ phase in the fused layer [15].

The LSM method produced the higher surface roughness of AZ80 magnesium alloys compared to MB26 due to the variation in cooling rate. A nanosecond pulsed fiber laser with the wavelength of 1060 nm is used for the LSM process. The process parameters such as pulse duration, repetition rate, and spot size are 220 ns, 500 kHz, and 44 μm, respectively. Alloys are irradiated with a laser power density of 1.20×10^7 W/cm^2 and at a scanning speed of 200 mm/s with 50% beam bath overlapping. The higher microhardness was observed for MB26 than the AZ80 due to the higher melting layer thickness [16]. The LSM is also used to study the grain size, microhardness of hybrid composites. The laser power is varied from 1.8 to 2.0 kW, the laser beam diameter range is 4.72–6.07 mm, standoff distance range is 35–45 mm and a constant scan speed of 400 mm/s is maintained. Argon shielding gas is used during the laser melting process to prevent the oxidation. The study found that the LSM treated hybrid metal matrix composite has lower grain size compared to untreated composites due to rapid solidification after LSM. The LSM produces the higher hardness of composites compared to untreated composite [17]. The effect of different laser power on microhardness and wear of AISI M2 high speed steel is studied by using LSM. The Nd: YAG laser, stand of distance varying from 1 to 2 cm, power varying from 600 to 1800 W, argon gas of 0.5 bar, laser spot varying from 2 to 4 mm and speed varying from 50 to 100 cm/min are used in this process. The results found that the maximum hardened depth of 0.85 mm is achieved by using power of 1400 W. The wear resistance of tool steel is nearly equal to conventionally hardened work material and it is shown in **Figure 10**. The reason for LSM produces high wear resistance and high hardened surface is due to the fine dendrites with dissolved carbides [18]. The LSM is also used to improve the hardness and wear resistance of Hastelloy C-276. The CW CO_2 laser with the parameters of 2 mm beam diameter, 0.6 MPa argon pressure, power varying from 1.25–1.75 kW, speed of 300 mm/min and interaction time of 400 ms are used in the work. The result found that the maximum hardness of 447 HV is achieved by using the power of 1.5 kW and scanning speed of 300 mm/min. The hardness is improved by 1.8 times compared to parent metal. The wear resistance of hastelloy is high in the sample laser treated at 1.5 kW of power and 300 mm/min speed and it is shown in **Figure 11**. This is due to the significant effect of grain refinement on hardness [19].

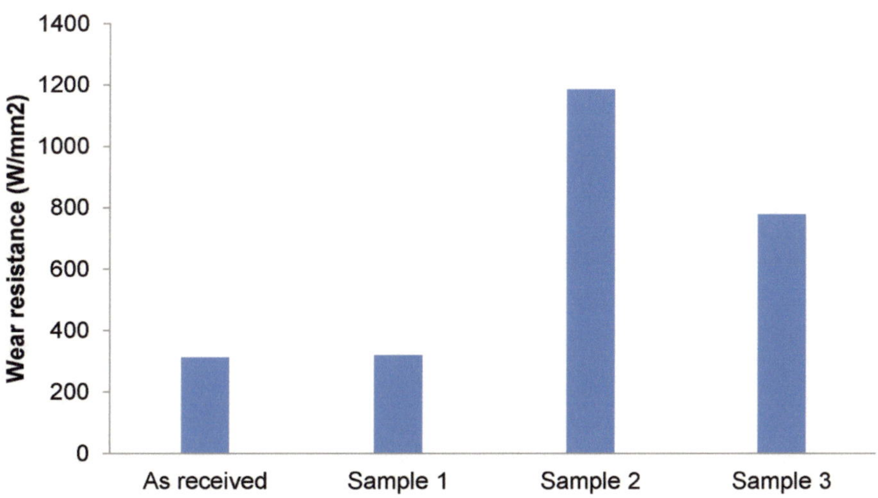

Figure 11.
The effect of LSM on wear resistance of Hastelloy C-276.

The laser surface melting is carried out on nodular cast iron (NCI) [20]. The laser parameters, power of 1.5 kW, scan speed of 600 mm/min, overlapping of 30% and defocus of 15 mm and argon gas are used to melt the NCI surface. The microstructure of as received nodular cast iron showed with more ferrite and less pearlite as shown in **Figure 12a**. The γ-phase dendrites and an interdendritic carbide structure were observed in the laser treated region and it is shown in **Figure 12b**. The reason for forming dendrite in the laser treated region is due to the rapid heating and solidification. The needle shape interdendritic structure of Fe_3C and M-phase is also observed due to the higher cooling rate. The convection is also the reason for forming of homogeneous dendritic. The small diameter of nodules is also observed in the bottom layer with partial dissolution of nodular graphite due to the heat treatment and self-quenching. The uneven martensite and dendrite phases are observed in the intermediate layer due to the rapid re-solidification of the melt pool. Finally, fine martensite is observed in the bottom region. Moreover, no cracks and no voids are observed in the processed depth.

The worn out surface of as received and laser melted surface is shown in **Figure 13a** and **b**. The LSM specimen wear track showed with smooth, minor grooves and delamination. The wear depth and pile-up of laser processed specimens are lesser than untreated specimens. The laser treated surfaces have

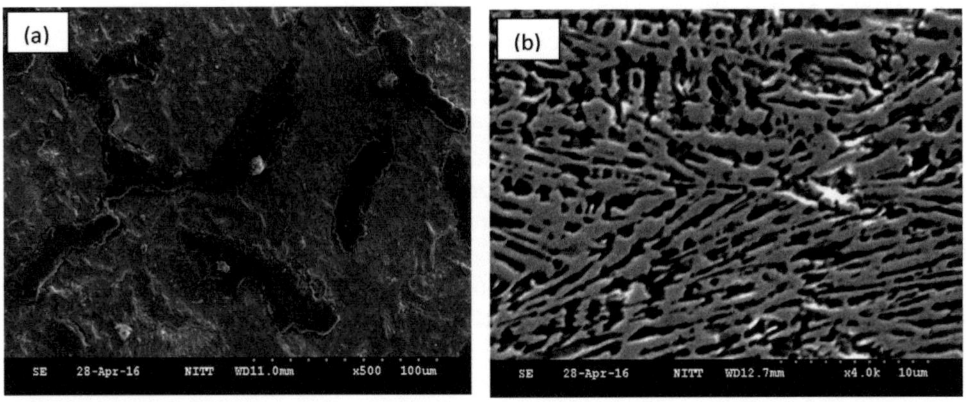

Figure 12.
Microstructure of; (a) as-received nodular cast iron, and (b) laser surface melted nodular cast iron.

Figure 13.
Worn out surface of; (a) base metal, (b) laser melted specimen.

fine grooves resulting in improving the wear resistance of specimens due to the microstructure changes. The root causes for improving the wear resistance of laser processed materials are fine M-phase and retained γ-phase with Fe_3C phase. The length of depth of hardness is increased by increasing the melted depth. The reasons are due to the precipitation hardening, residual stress by refinement of grains through rapid re-solidification. The cooling rate and thermal gradient also support the refinement of grains resulting in increased the hardness of the laser treated zone. Compared to hardness of substrate material, the laser processed depth has four time higher hardness due to the uniform grain structure. The partially melted zone shows the higher hardness due to the graphite nodules and fine ledeburite microstructure with the graphite interface. The wear loss is calculated for both the laser processed sample and untreated sample. The laser processed samples showed less wear than substrate.

4. Laser surface alloying (LSA)

Laser surface alloying is a material processing technique that utilizes the focused laser sources and produces the high power density to melt the metal coating and a portion of the underlying substrate. The schematic view of laser surface alloying is shown in **Figure 14**. The schematic diagram of shape and dimensions of laser surface alloyed zone is shown in **Figure 15**. Here, W = width, T = thickness, B = build-up and D = melted depth. Aluminum alloys are widely used in automobile and aerospace applications due to the availability and low cost, ductility, good strength-to-weight ratio and lightweight. These alloys have low hardness and poor tribological properties which leads to wear problem. Hence, the additional protection is required to enhance the wear resistance properties to localized areas. So, LSA can be used to improve the surface properties of aluminum alloys, titanium alloys, magnesium alloys, copper alloys and nickel-copper alloys. The laser alloyed component properties are depending upon the selection of alloy material, composition and elemental surface distribution. These factors are affecting the microstructural

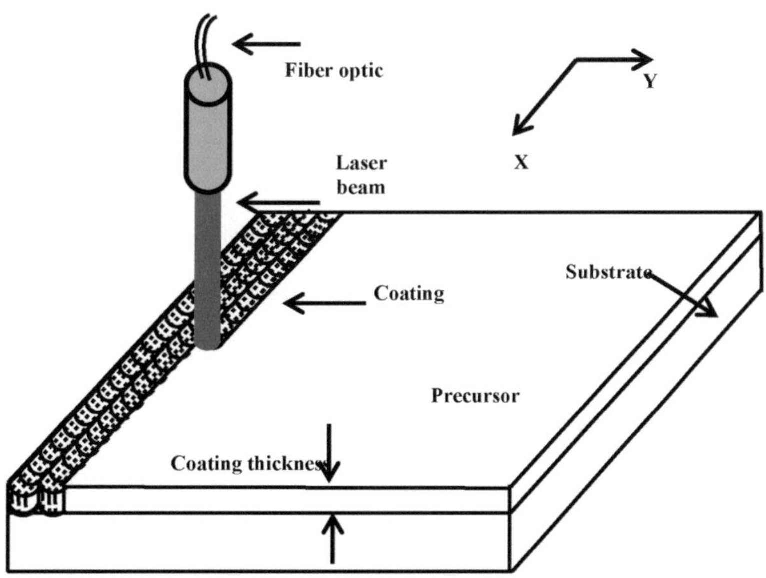

Figure 14.
Schematic view of laser surface alloying.

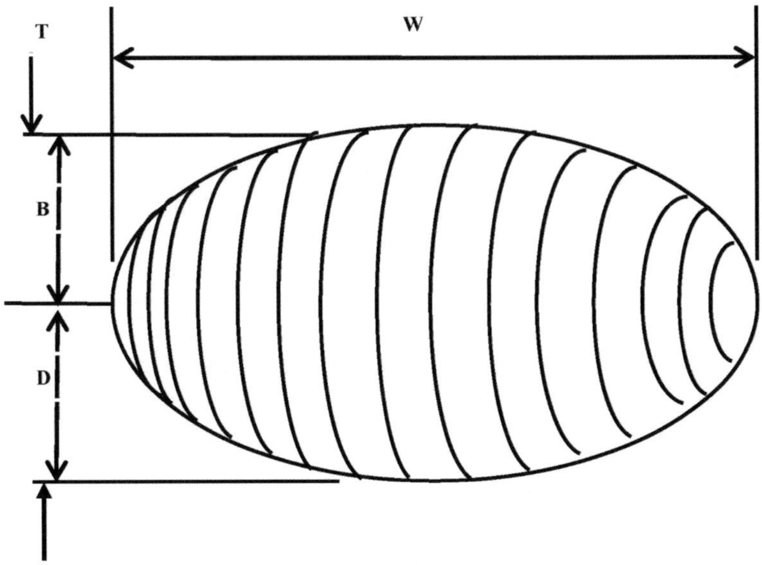

Figure 15.
Schematic diagram of shape and dimensions of laser surface alloyed zone.

development in the alloyed surface. The ceramic alloys, carbide, oxide and boride (SiC, WC, TiO_2, TiB_2 and TiC) are widely used as the coating material on aluminum alloys due to the low density, high hardness, good wear, high melting temperature and corrosion resistance. The hybrid ceramics, a component coating on aluminum produces better wear resistance than the single ceramics component coating. Titanium is added to the carbon resulting in forming TiC to improve the surface properties and by the same to prevent the formation of Al_4C_3 carbides. A study on $FeCoCrAlCuNi_x$ high entropy alloy coating on pure copper is carried out using LSA to evaluate the microhardness and wear. The laser power 1.7 kW, laser spot diameter 1.2 mm, scanning speed 2.0–3.0 mm/s, argon as shielding gas and flow rate 12 L/min are used in this process.

Figure 16a shows the microstructure of HEA $FeCoCrAlCuNi_x$. The HEA coating have high density, little holes and adequate metallurgical bonds to substrate. It is noticed that the dilution ratio of the tested HEA coating is higher than 20%. Typical dendrite and interdendrite structures are clearly observed in Ni05 and Ni10 HEAs (**Figure 16b** and **c**), while only one phase was observed for Ni15 HEA (**Figure 16d**). Compared to hardness of copper, coated copper produces higher hardness and it is shown in **Figure 17** [21]. The effect of addition of Ni–Cr–Si–B alloy to brass substrate was studied through LSA. The 2 kW CW Nd-YAG laser with a spot diameter of 3 mm, the laser power density varied between 141 and 212 W/mm, while the scanning speed is kept constant at 5 mm/s. Argon with a flow rate of 15 l/min is used as the shielding gas to prevent the oxidation. Laser surfacing is achieved by overlapping of adjacent tracks, with an overlapping ratio of 50%.The hardness of the modified layers increased slightly from the surface to a maximum and sharply fell to the value of the substrate at the interface between the treated layer and the substrate. The increases in hardness observed for the modified layer is attributed to the formation of hard borides [22]. The effects of addition of SiC and TiO_2 to aluminum alloy are studied by continuous mode CO_2 laser. The CO_2 laser with the parameters of 1.7 kW, scan speed of 400 mm/min, standoff distance of 40 mm and laser beam diameter of 7.4 mm are used for SiC alloying. The CO_2 laser with the parameters of 1.8 kW, scan speed of 300 mm/min, standoff distance of 30 mm and laser beam diameter of 5.8 mm are used for TiO_2 alloying. The result found that the ceramic nature of SiC

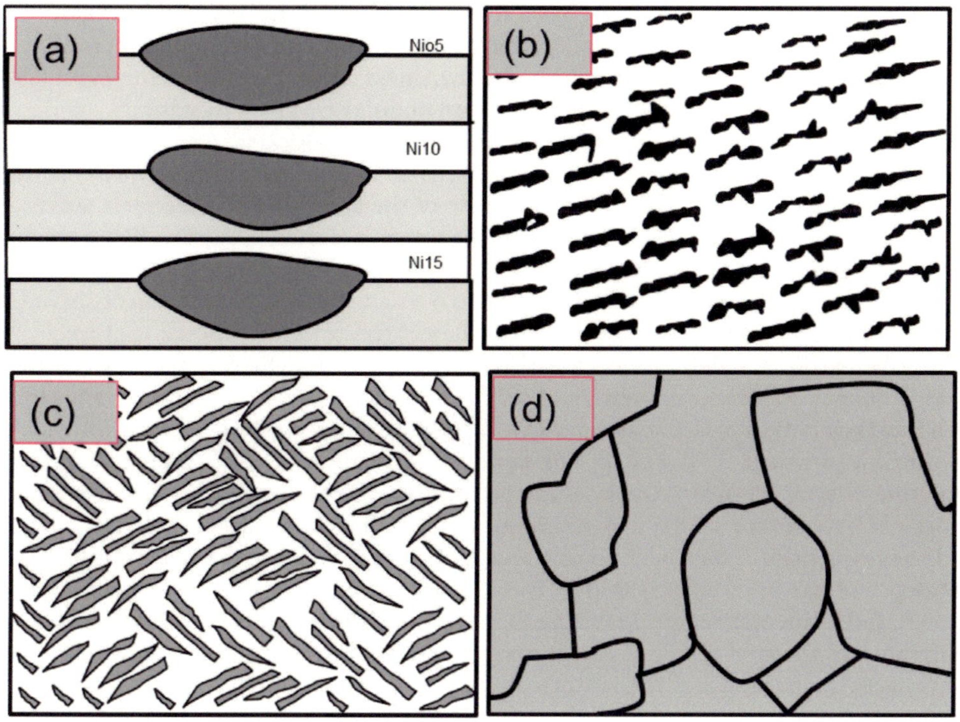

Figure 16.
Microstructure images of (a) FeCoCrAlCuNix HEA coatings on cross sectional view, (b) high magnifications image of Ni05 HEA (c), Ni10 HEA (d) and Ni15 HEA.

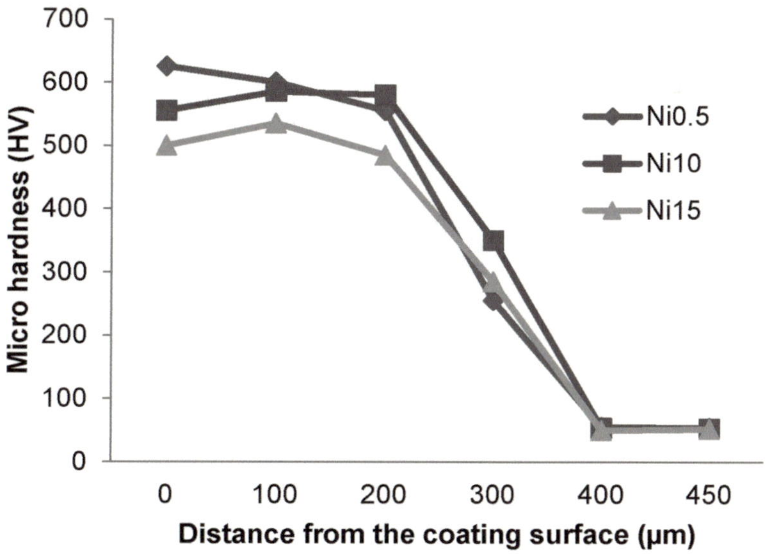

Figure 17.
Microhardness of $FeCoCrAlCuNi_x$ HEA coatings.

and TiO_2 improved microhardness of alloyed zone from 30 $HV_{0.3}$ substrate material to 180 $HV_{0.3}$ with SiC and 220 $HV_{0.3}$ with TiO_2 [23].

A study on the effect of addition of WC + Co + NiCr to AISI 304 stainless steel through Nd: YAG laser. The 5 kW Nd: YAG with beam diameter of 4 mm, power varying from 1 to 3 kW, scan speed from 0.005–0.1 m/s and argon gas of 5 L/min are used in the alloying process. The experimental result found that the LSA has been performed to form a defect free and uniform alloy zone. Compared to hardness of substrate, laser alloying produces the higher hardness due to the grain refinement [24].

The laser surface alloying is carried out on nodular cast iron by adding Ni-20%Cr alloy [20]. The laser parameters, power of 1.5 kW, scan speed of 600 mm/min, overlapping of 30% and defocus of 15 mm and argon gas are used to alloying the NCI surface. The microstructure of the laser alloyed specimen, worn out surface of substrate and laser alloyed specimen is shown in **Figure 18a–c** respectively. The ledeburite and pre-eutectic austenite are observed in the LSA surface. In addition, γ-phase (austenite) to M-phase (martensite) is transformation observed. The laser alloyed surface has produced the defect free and fine microstructure. The γ-phase has a higher percentage of Ni than cementite, whereas the Fe3C phase has Cr more and Ni less element. Hence, the presence of Fe3C on the laser-alloyed surface is rich in Cr and the γ-phase was supported through the solid solution of both alloy powders of Ni and Cr. The rapid solidification is the reason for obtaining the fine microstructure in the laser alloyed surface. The laser processed worn out surfaces have severe plastic deformation, wear track, delamination, grooves and adhesive particles. The NiCr alloying is also observed by using the LSA. The length of depth of hardness is increased by increasing the melted depth. The reasons are due to the refinement of grains through rapid re-solidification. The rate of cooling rate and thermal gradient also support the refinement of grains resulting in increased the hardness of the laser treated zone. Compared to hardness of substrate material, the laser processed depth has 2.62 time higher hardness due to the uniform grain structure. The wear loss is calculated for laser processed sample and untreated

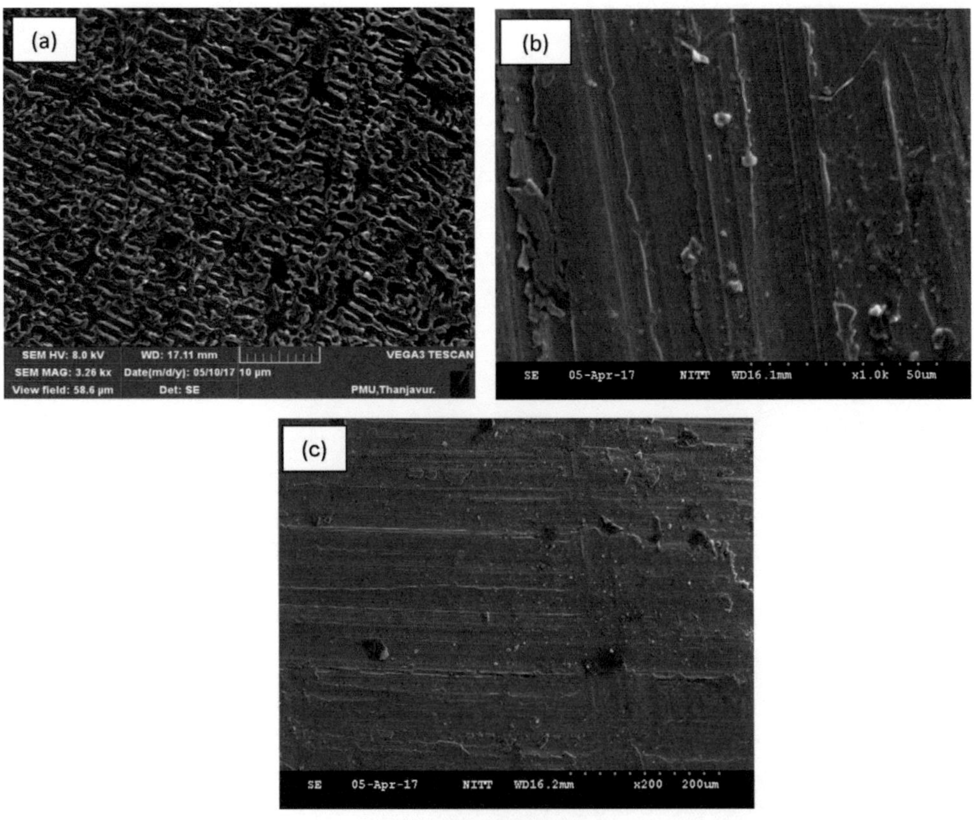

Figure 18.
Microstructure of LSA specimen (a), worn out surface of substrate (b), and worn out substrate of LSA (c).

sample. The laser processed samples are produced lesser wear rate than substrate due to the improved hardness.

5. Laser cladding (LC)

Laser cladding is similar to arc welding. The laser is used to melt the clad material coated on the substrate. The powder, wire and strip form of clad materials are commercially available to perform by different laser processes. The major benefits of LC have low porosity, good surface uniformity and low dilution. The clad materials have rapid quench and cooling down after deposition resulted in a fine grained microstructure. The laser is used to deposit clad material on substrate through the interaction of powder with laser. The substrate permits the melt pool to solidify and form the solid track. The schematic of laser cladding process is shown in **Figure 19**. Compared to other different surface processing used to enhance the wear and corrosion resistance of substrate, LC is an attractive alternative method. This is due to the intrinsic properties of laser radiation. The LC benefits are high input energy, low distortion, and minimum dilutions observed between the substrate, processing flexibility and cladding on small areas. The LC can be used in surface alloys and composites in order to achieve the required properties. The LC produces desired properties are obtained by varying the process parameters such as laser beam power density, laser beam diameter at the workpiece surface and laser beam travel speed.

The laser solution strengthening, laser surface alloying and laser cladding have highly correlation to corrosion and erosion resistance. The laser solution strengthening and laser surface alloying are used to improve the erosion and corrosion resistance of old components without changing their sizes whereas laser cladding is used to repair wasted components by restoring their size. The high entropy alloy of $CoCrFeNiNb_x$ is coated to a pure titanium sheet by using laser cladding to study the hardness of the material. The laser cladding parameters such as power of 100 W, scanning speed of 8 mm/s, defocusing amount of +2 mm, pulse duration of 5 ms,

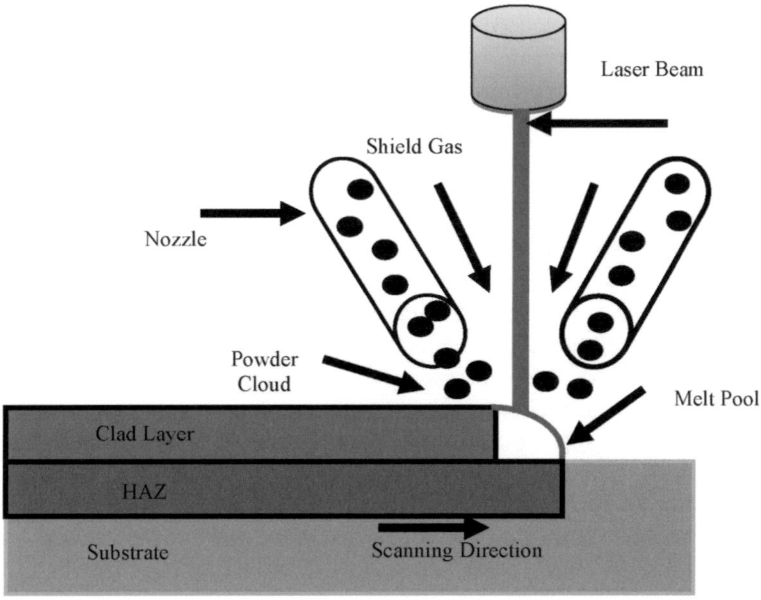

Figure 19.
Schematic of laser cladding process.

frequency of 20 Hz, beam diameter of 1 mm power density of 127.4 W/mm^2 and linear energy density of 12.5 J/mm are used in this process. The result found that the CoCrFeNiNbx HEA coated on titanium sheet produces higher hardness compared to the pure titanium. The Nb coating produces significant improvement in hardness compared to pure titanium due to the consisted phase of BCC solid solution with equiaxed bulk grain morphology and Cr_2Ti Laves phase [25]. A comparison is performed between the thermal spray coating and laser cladding performance on steel. The laser cladding conditions, power of 780 W, cladding speed of 4.3 mm/s, powder feed rate of 6 g/min and argon gas are used. The thermal spray conditions, distance of 200 mm, acetylene (0.7 bar) and oxygen (4 bar) gas are used. The Metco 15E powder is used in both the processes. The result found that the cladded layer produced the high hardness, crack free, and good adherence to substrate whereas flame coating produces high porosity, minimum dilution and oxides inclusions [26]. The Inconel 625 coating performance on steel is evaluated by arc welding and laser cladding based on the microstructure, wear resistance and hardness. The parameters, power of 1200 W, scan speed of 2 mm/s, powder feed rate of 5 g/min, shielding gas flow rate of 5 L/min and powder feeding gas flow rate of 8 L/min are used. The result found that the arc welded and laser cladded Inconel 625 coatings have Ni (fcc) solid solution phase, and fine microstructure. The arc welded coating to Inconel 625 is produced slightly lower hardness compared to laser cladding coating. This is due to the microstructure developed in the arc welding. The laser cladded Inconel 625 coating is preferred due to its better mechanical performance such as hardness and wear resistance at both room and elevated temperature [27]. The 316 stainless steel powders coated on EN3 mild steel is to evaluate clad geometry and distribution of elements by laser cladding. The 2 kW continuous wave CO_2 with laser power 1.8 kW, beam spot diameter 2–5 mm, powder feed rate 0.160–0.220 g/s, substrate traverse speed 7–40 mm/s are used. The stainless steel powder coating provides the sound coating and no porosity [28]. The Fe-Cr-Si-B alloy powder coating is performed on low carbon steel using laser cladding to evaluate the microstructure, hardness, wear resistance and corrosion resistance. The result identified that the Fe-Cr-Si-B alloy powder coating provides higher wear resistance, high hardness and high corrosion resistance compared to substrate [29]. The CPM 15 V, CPM 10 V, CPM 9 V, D2 and M4 coatings are provided on AISI 1070 carbon steel by laser cladding. The laser cladding conditions, power varying from 2.5–2.75 W, laser beam diameter varying from 2 mm, substrate traverse speed varying from 7.6–8.6 mm/s, powder feed rate varying from 20 to 9 g/min and overlap varying from 30 to 50% are used in the process. The abrasive wear resistance of the laser-clad CPM 15 V and CPM 10 V coatings is superior performance than D2 steel, whereas the wear resistance of the CPM 9 V and M4 coatings is inferior to that of the D2 [30].

Figure 20a shows the microstructure of Colmonoy 6 cladding on Inconel 625 [31]. The laser cladding parameters are 400 mm/min speed, feed rate of 4 g/min, power of 1000 W, argon pressure of 1 bar with flow rate of 25 lpm and 150 degrees preheating used in this process. The clad surfaces have no defects, uniform dendrite eutectic phases observed. There are two regions represented in the cladded surface such as darker region for boride content and lighter region for γ-nickel. The high quantity of intermetallic lave phase is observed in the cladded surface. **Figure 20b** shows the worn out surface of substrate Inconel 625. Compared to wear intensity of sample, laser cladded surfaces have lesser wear. The plow marks are also observed in the worn out surface substrate due to the less wear resistance and high plastic deformation. The higher material removal rate of the sample is observed than the cladded surface. **Figure 20c** shows the laser cladded worn out surface. The few debris particles, few depth of wear track and few grooves are observed in the cladded surface. This is due to the high hardness of the clad layer. Therefore, better

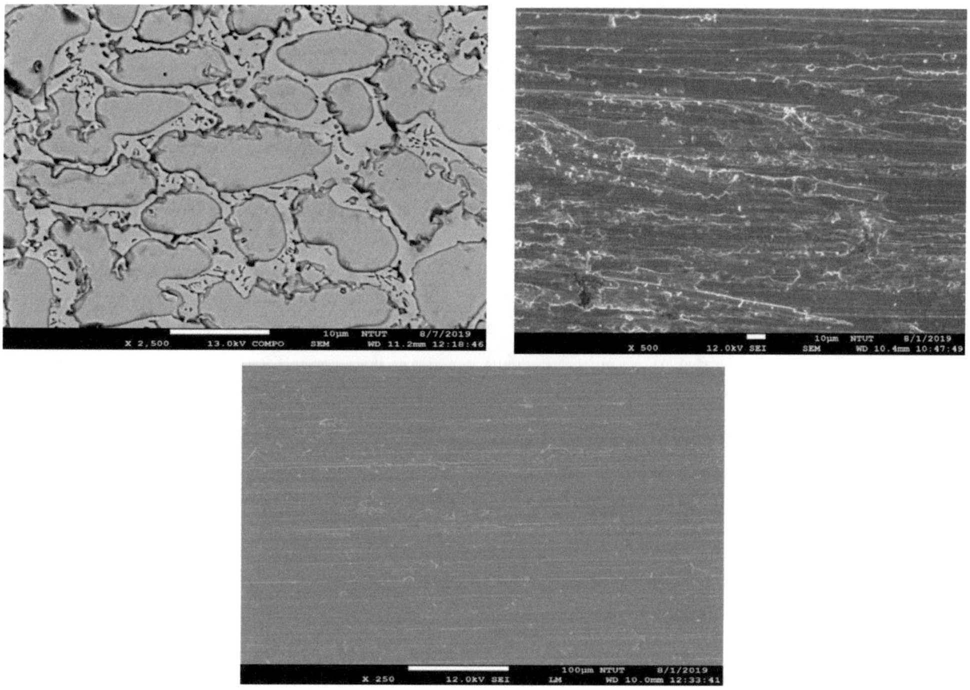

Figure 20.
Microstructure of Colmonoy 6 clad (a), worn out surface of substrate (b) and worn out surface of laser cladding (c).

protection is provided by the clad layer over the untreated surface. The more hardness is observed in the clad surface than the base metal. The reasons for increasing the hardness of clad surface is due to the defect free cladding, proper fusion and laves phase presented. The reason for decreasing the hardness of base material is due to weak intermetallic phases.

The coefficient of friction (CoF) and wear behavior of coated and substrate found that the CoF is increased with increased sliding distance due to the reduced adhesion resistance and increasing heat between points of contact. The more CoF is observed in the substrate sample than clad sample due to the adhesion effect. The less CoF is observed in the clad sample due to the hard laves phases. It is found that low mass loss is observed in the clad sample compared to base material. The wear loss is highly related to the hardness and base material produces poor wear resistance when compared to clad surface.

6. Laser texturing

Laser texturing is a process that alters a material surface property by modifying its texture and roughness. The laser beam creates micro patterns on the surface through laser ablation, removing layers with micrometer precision and perfect repeatability. Typical patterns include dimples, grooves, and free forms. Laser surface texturing can be used to improve properties like adherence, wettability, electrical and thermal conductivity, and friction. For example, the method can increase surface adherence before applying common coatings like adhesives, paint or ceramic. Laser texturing can also be used to prepare surfaces for thermal spray coating and laser cladding as well as to improve the performance of mechanical seals. Surface treatments like abrasive blasting and chemical etching processes need

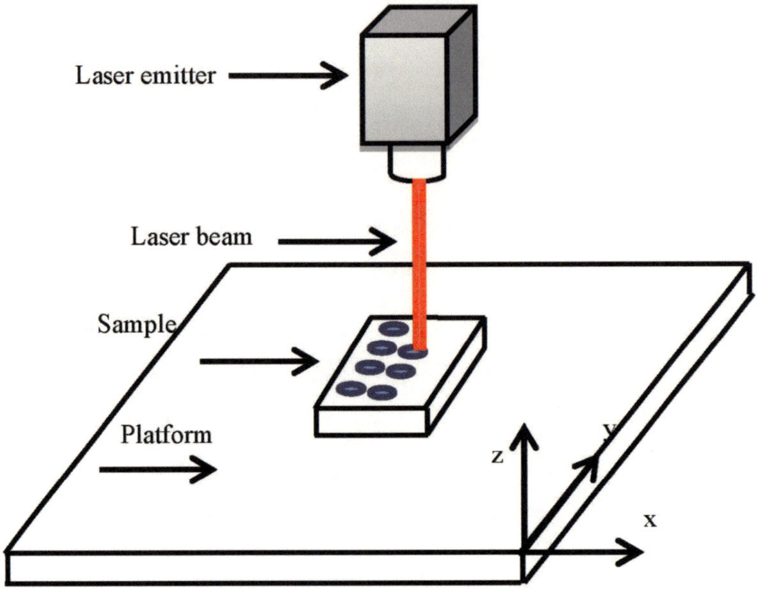

Figure 21.
Schematic of laser texturing process.

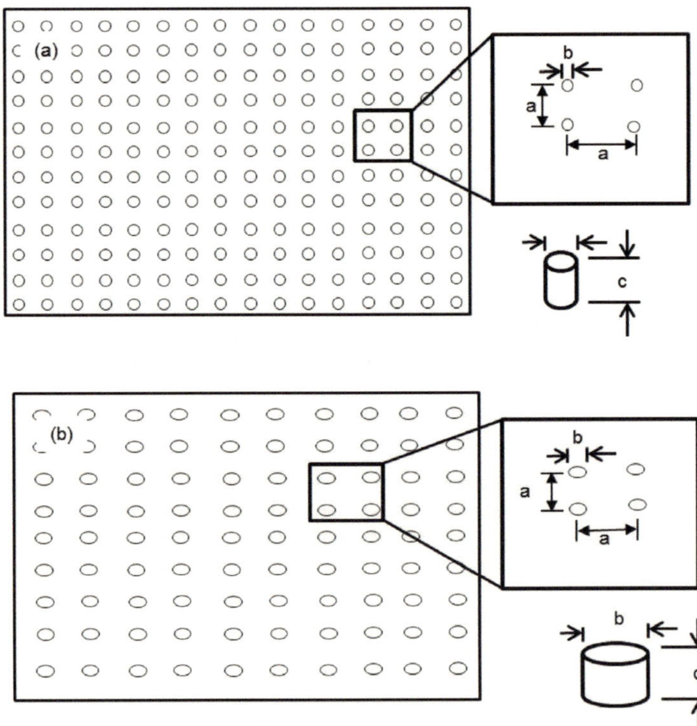

Figure 22.
Schematic of laser surface texture dimensions: (a) circle, (b) oval.

consumables like steel grits and acid to texture surfaces. Unlike those treatments, the laser texturing process functions without consumables. This results in low operating costs, low maintenance, and improved health and safety in the workplace.

Operators will not need to handle chemicals, wear protective equipment, and stop operations to replace consumables. Laser texturing uses laser ablation to selectively remove materials from specific surface areas. By adjusting the laser's parameters, the surface is removed as well as creates different patterns. This typically increases roughness, creating surface textures that can easily lodge adhesives and provide additional anchoring surface. To reach the material's ablation threshold, pulsed lasers concentrate energy to reach a high peak power. Typically, the pulse duration is 100 nanoseconds, and each pulse contains between 0.5 and 1 millijoules. The time required to texture a surface depends on the material, the desired roughness level, and the laser system's output power. The application of laser texturing is in adhesive bonding, mechanical seals, painting and coatings. The laser texturing process is shown in **Figure 21**. The circle and oval shape dimple texturing on metal can be made using a laser and the schematic diagram is shown in **Figure 22a–b**. Where, a = pitch, b = diameter, and c = height.

7. Conclusion

The new materials have been developed every day to meet the demand of competitive situations. The surface properties of substrate can be improved by a number of methods such as laser surface alterations such as surface hardening, melting, alloying, cladding and texturing in order to improve the mechanical performance and tribological behavior. In this work, the effect of laser process parameters on microstructure, hardness and wear rate of materials have been presented. The laser surface hardening is needed to high stressed components namely gear teeth, gears, shafts, camshafts, axles, cylinder liners and exhaust valves. The laser surface melting can be adopted in biomedical alloys, sport cars and power plants made of stainless steel, magnesium alloy and superalloys. The locomotive, aerospace and structural components made of aluminum alloys, titanium alloys and magnesium alloys have required the laser surface alloying to improve the surface properties of metals. The repaired and refurbishment components such as internal combustion engine parts, gas turbine, turbine blades and tools are highly needed the laser surface cladding to improve the surface properties of metals. The texturing on material is used to increase the tribological characteristics of materials resulting in improved surface roughness, wettability, improve load capacity, wear rates, lubricating lifetime and reduce friction coefficient. Hence, the laser based surface modification techniques can be adopted to improve the performance of the components.

Acknowledgements

The authors wish to thank the Ministry of Science and Technology, Taiwan ROC for the financial support to carry out this work.

Conflict of interest

The authors declare that they have no conflicts of interest in the work.

Author details

Natarajan Jeyaprakash[1*], Che-Hua Yang[1,2] and Durairaj Raj Kumar[3]

1 Additive Manufacturing Center for Mass Customization Production, National Taipei University of Technology, Taipei, Taiwan, ROC

2 Institute of Manufacturing Technology, National Taipei University of Technology, Taipei, Taiwan, ROC

3 Department of Mechanical Engineering, MAM School of Engineering, Tiruchirappalli, India

*Address all correspondence to: prakash84gct@gmail.com; prakash@ntut.edu.tw

References

[1] Totten, G.E. (2006) Steel Heat Treatment: Equipment and Process Design Handbook, 2nd ed., Taylor and Francis Group.

[2] Tani, G., Orazi, L. and Fortunato, A. (2008) 'Prediction of hypo eutectoid steel softening due to tempering phenomena in laser surface hardening', CIRP Annals-Manufacturing Technology, Vol. 57, pp.209-212.

[3] Carrera-Espinoza, R., Rojo Valerio, A., del Prado Villasana, J., Yescas Hernández, J.A., Moreno-Garibaldi, P., Cruz-Gómez, M.A. and Figueroa López, U. (2020) 'Surface Laser Quenching as an Alternative Method for Conventional Quenching and Tempering Treatment of 1538 MV Steel', Advances in Materials Science and Engineering.

[4] Liu, Q., Song, Y., Yang, Y., Xu, G. and Zhao, Z. (1998) 'On the laser quenching of the groove of the piston head in large diesel engines', Journal of materials engineering and performance, Vol. 7, pp.402-406.

[5] Wang, B., Pan, Y., Liu, Y., Lyu, N., Barber, G.C., Wang, R., Cui, W., Qiu, F. and Hu, M. (2020) 'Effects of quench-tempering and laser hardening treatment on wear resistance of gray cast iron', Journal of Materials Research and Technology, Vol. 9, pp. 8163-8171.

[6] Slatter, T., Taylor, H., Lewis, R. and King, P. (2009) 'The influence of laser hardening on wear in the valve and valve seat contact', Wear, Vol. 267, pp.797-806.

[7] Pashby, I.R., Barnes, S. and Bryden, B.G. (2003) 'Surface hardening of steel using a high power diode laser', Journal of Materials Processing Technology, Vol. 139, pp.585-588.

[8] SeDao, Hua, M., Shao, T.M. and Tam, H.Y. (2009) 'Surface modification of DF-2 tool steel under the scan of a YAG laser in continuously moving mode', Journal of Materials Processing Technology, Vol. 209, pp.4689-4697.

[9] Rana, J., Goswami, G.L., Jha, S.K. Mishra, P.K. and Prasad, B.V.S.S.S. (2007) 'Experimental studies on the microstructure and hardness of laser-treated steel specimens', Optics and Laser Technology, Vol. 39, pp.385-393.

[10] Barka, N., Sattarpanah Karganroudi, S., Fakir, R., Thibeault, P. and Feujofack Kemda, V.B., 2020. Effects of Laser Hardening Process Parameters on Hardness Profile of 4340 Steel Spline—An Experimental Approach. Coatings, 10(4), p.342.

[11] Zhou, W., Zhou, Y., Wu, N. and Maharjan, N., 2019. Laser surface hardening of AISI 1055 steel in water submerged condition.

[12] Zhou, J., Xu, J., Huang, S., Hu, Z., Meng, X. and Feng, X., 2017. Effect of laser surface melting with alternating magnetic field on wear and corrosion resistance of magnesium alloy. Surface and Coatings Technology, 309, pp.212-219.

[13] Majumdar, J.D., Galun, R., Mordike, B.L. and Manna, I., 2003. Effect of laser surface melting on corrosion and wear resistance of a commercial magnesium alloy. Materials Science and Engineering: A, 361(1-2), pp.119-129.

[14] Zhang, L., Yu, G., Li, S., He, X., Xie, X., Xia, C., Ning, W. and Zheng, C., 2019. The effect of laser surface melting on grain refinement of phase separated Cu-Cr alloy. Optics & Laser Technology, 119, p.105577.

[15] Cui, Y., 2018. Influence of laser surface melting on tribological behaviour of AZ31B. Surface Engineering, 34(4), pp.296-300.

[16] Li, Y., Arthanari, S. and Guan, Y., 2019. Influence of laser surface melting on the properties of MB26 and AZ80 magnesium alloys. Surface and Coatings Technology, 378, p.124964.

[17] Bannaravuri, P.K., Birru, A.K. and Dixit, U.S., 2020. Effect of laser surface melting on surface integrity of Al– 4.5 Cu composites reinforced with SiC and MoS2. Transactions of Nonferrous Metals Society of China, 30(2), pp.344-362.

[18] Newishy, M., Abdel-Aleem, H., Elkousy, M.R., El-Mahallawi, I. and El-Batahgy, A., 2018. Surface Treatment of AISI M2 Tool Steel by Laser Melting. In Key Engineering Materials (Vol. 786, pp. 128-133). Trans Tech Publications Ltd.

[19] Hashim, M. and Duraiselvam, M., 2015. Enhancing tribological and corrosion resistance of Hastelloy C-276 through laser surface treatment. In Materials Science Forum (Vol. 830, pp. 659-662). Trans Tech Publications Ltd.

[20] Jeyaprakash, N., Yang, C.H., Duraiselvam, M. and Sivasankaran, S., 2020. Comparative study of laser melting and pre-placed Ni–20% Cr alloying over nodular iron surface. Archives of Civil and Mechanical Engineering, 20(1), pp.1-12.

[21] Wu, C.L., Zhang, S., Zhang, C.H., Zhang, H. and Dong, S.Y., 2017. Phase evolution and properties in laser surface alloying of FeCoCrAlCuNix high-entropy alloy on copper substrate. Surface and Coatings Technology, 315, pp.368-376.

[22] Tam, K.F., Cheng, F.T. and Man, H.C., 2002. Enhancement of cavitation erosion and corrosion resistance of brass by laser surface alloying with Ni–Cr–Si–B. Surface and Coatings Technology, 149(1), pp.36-44.

[23] Jiru, W.G., Sankar, M.R. and Dixit, U.S., 2017. Investigation of microstructure and microhardness in laser surface alloyed aluminium with TiO2 and SiC powders. Materials Today: Proceedings, 4(2), pp.717-724.

[24] Chakraborty, A., Pityana, S.L. and Dutta Majumdar, J., 2017. Laser surface alloying of AISI 304 stainless steel with WC+ Co+ NiCr for improving wear resistance, Procedia Manufacturing, 7, pp. 8-14.

[25] Xiang, K., Chen, L.Y., Chai, L., Guo, N. and Wang, H., 2020. Microstructural characteristics and properties of CoCrFeNiNbx high-entropy alloy coatings on pure titanium substrate by pulsed laser cladding. Applied Surface Science, p.146214.

[26] Pascu, A., Hulka, I., Tierean, M.H., Croitoru, C., Stanciu, E.M. and Roată, I.C., 2016. A comparison of flame coating and laser cladding using Ni based powders. In Solid State Phenomena (Vol. 254, pp. 77-82). Trans Tech Publications Ltd.

[27] Feng, K., Chen, Y., Deng, P., Li, Y., Zhao, H., Lu, F., Li, R., Huang, J. and Li, Z., 2017. Improved high-temperature hardness and wear resistance of Inconel 625 coatings fabricated by laser cladding. Journal of Materials Processing Technology, 243, pp.82-91.

[28] Weerasinghe, V.M., Steen, W.M. and West, D.R.F., 1987. Laser deposited austenitic stainless steel clad layers. Surface engineering, 3(2), pp.147-153.

[29] Yang, Yan CJ and Wang 1987 Laser cladding of FeCrSiB alloy Lasers 14 548-52

[30] Wang, S.H., Chen, J.Y. and Xue, L., 2006. A study of the abrasive wear behaviour of laser-clad tool steel coatings. Surface and Coatings Technology, 200(11), pp.3446-3458.

[31] Jeyaprakash, N., Yang, C.H. and Ramkumar, K.R., 2020. Microstructure and wear resistance of laser cladded Inconel 625 and Colmonoy 6 depositions on Inconel 625 substrate. Applied Physics A, 126, pp.1-11.

Chapter 5

Laser Machining

Natarajan Jeyaprakash, Che-Hua Yang
and Manickam Bhuvanesh Kumar

Abstract

The increasing demands of materials with superior properties are given priority by most of the industries in recent years due to their higher performance levels. Machining of hard materials is a challenging task since it involves higher cutting forces and rapid tool wear. This leads to complexity in shaping these difficult-to machine materials such as advanced composite and ceramics. There have been many alternative techniques developed to overcome the shortcomings of conventional machining processes. Laser beam machining (LBM) is one of the advanced non-contact machining processes that employ monochromatic light with high frequency for machining using thermal energy. The highly energized photos are focused on a material cause heating, melting and vaporizes the material which is effectively used to remove unwanted portion of a material. Due to higher coherency of laser beam, materials can be machined very precisely than conventional machining processes. Generally, the laser-based material processing is suitable for a brittle type of material with minimum conductivity. However, this laser machining can be used for all kinds of materials in most cases. This chapter provides the principle of laser and its types, mechanism of material removal using laser, applications, advantages, and limitations of LBM.

Keywords: laser, monochromatic, machining, laser ablation, stimulated emission

1. Introduction

The growing product development for advanced applications such as aerospace, automobiles, electronics and medical devices requires materials with high strength-to-weight ratio. Advanced materials with superior properties are being developed by researchers around the world for meeting the growing demand. The materials such as nickel, titanium and their alloys, ceramics are known not only for high strength-to-weight ratio but also for higher level of corrosion resistance, prolonging capacity at higher temperatures with superior mechanical strength comparing to other engineering materials [1]. These materials have greater properties such as higher density and melting point, ductile, higher hardness and strength, hence conventionally machining these materials is very challenging. Despite, it can be machined using conventional techniques, but higher cutting forces and rigorous tool wear attributes to huge cost in shaping these materials to the requirement. Hence there were many unconventional machining processes (UMPs) developed to replace conventional machining processes. One of the UMPs is laser beam machining (LBM) which is extensively used machining those difficult-to-machine materials. LBM is considered suitable for machining hard materials LBM is characterized by independency to hardness property of work material. LBM is gaining attention among the researchers and industry people because of its advantages such as higher light intensity with low power requirement, good focusing property within short duration of pulse, uniform heat distribution, eco-friendly nature which results in accuracy in machining, narrow heat affected zone, increased productivity and reduced manufacturing cost [2]. The upcoming sections describes in detail about the principle of laser and its types, mechanism of laser machining, advantages, applications and limitations of using LBM.

2. Principle of laser

The principle of Light amplification by stimulated emission of radiation (LASER) was first hypothesized by Albert Einstein in the year of 1917 but it took almost half a century to construct a working laser. Around 1960, a first experimental setup of working industrial laser is developed. In many cases, the laser is different from the normal light in a way that it carries photons of higher frequencies. However, in some cases, the infrared laser has photons with low frequency than normal light. The frequencies of all the photons contained in a laser light are all same hence laser is characterized by coherence. The photons carried by a light can stimulate the electrons in an atom therefore it emits same frequency photons [3]. Based on this principle, laser produces high energy coherent light. Since laser is the fundamental part of any laser-based system, it is essential to understand the principle of laser light production.

Stimulation and amplification is the process (called as lasing) by which laser system converts electrical energy into a light of high intensity energy. The medium by which the lasing process carried out is called lasing medium. In any model of an atom, positively charged nucleus is surrounded by negatively charged electrons rotating at some specified path called orbits. The diameter and geometry of the orbit vary based on many parameters including number of electrons, surrounded magnetic field, structure of electrons and the existence of neighbor atoms. Every electron presents in the orbital connected with a distinctive energy level. An atom is said to be at ground level when it is at absolute zero temperature in which all the electrons reside in their lowest potential energy. Energy from any exciting sources such as electronic pulsation at higher temperature, chemical reaction or photon can be absorbed by an electron at ground level. After absorbing the energy, it excites to a higher energy level as schematically shown in **Figure 1**. Thus the movement of electron from lower to higher energy level is accomplished. Upon reaching higher energy levels, electron attains an unstable energy band. Immediately within very short time (tens of nanosecond) it starts moving back to ground state by releasing

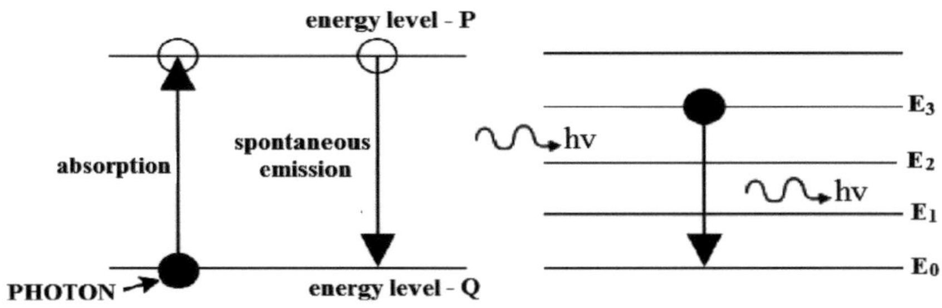

Figure 1.
Excitation between energy levels.

a photon and this process is termed as "spontaneous emission." The frequency of emitted photon would be equal to the frequency of exciting photon.

Sometime, when the electrons put into a meta-stable band due to energy change, the electron stays in the higher energy level itself for a short time (micro to milliseconds). The state by which more number of electrons stays in meta-stable energy level compared to the atoms in the ground level of a material is called "population inversion." These electrons are stimulated by suitable energy or frequency photons to come back to ground state. Photons emission due to this stimulated return of electrons is termed as "stimulated emission." In this way, the emitted photons along with one original photon temporarily having some spatial phase would create coherent laser beam. From the schematic representation of stimulated absorption, spontaneous emission, and stimulated emission as shown in **Figure 2**, position of electrons in various energy levels are shown.

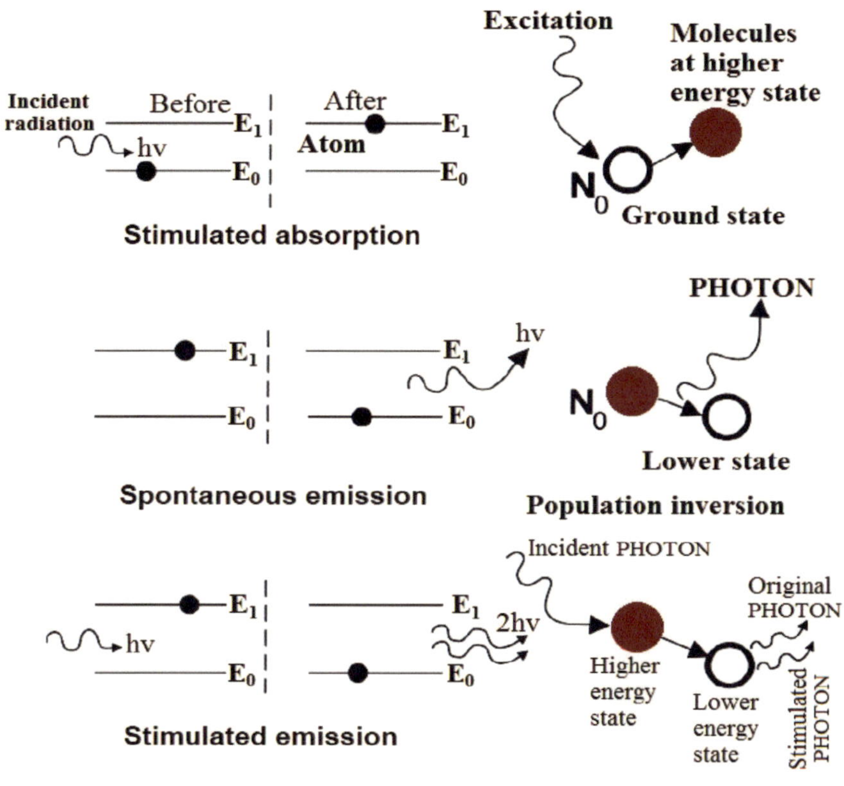

Figure 2.
Excitation between energy levels.

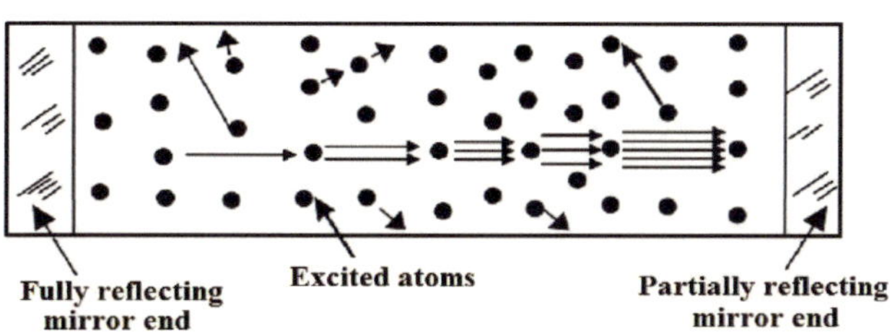

Figure 3.
Excitation between energy levels.

The working of laser is schematically represented in **Figure 3**. A lasing medium contained by a cylindrical glass container is closed using completely (100%) reflecting mirror on one end and partially reflecting mirror on the other end. When the glass vessel is exposed to a light using flash lamps, the photons of light excites the atoms of lasing medium thus population inversion is obtained. Further due to stimulated emission, photons are emitted. These stimulated photons in the longitudinal direction form a high intense, coherent and highly directional laser beam. Most of the stimulated photons would not be in the longitudinal direction and these photons usually generate waste heat and finally lost.

3. Properties of laser

The distinctive properties of laser are coherence, highly monochromatic, intensive radiance and directionality. These optical properties can be quantified for analyzing the laser properties.

3.1 Coherence

The relationship between magnetic and electronic components of electromagnetic wave refers to coherence property. The light beam is said to be coherent when these components are properly aligned as shown in **Figure 4**. There are two terms of coherence for a laser as spatial coherence and temporal coherence. Coherence is said to be spatial when the correlation of phases happens at different points in a space at a single point of time whereas in temporal coherence, correlation happens at single point in a space over a time period. **Figure 5** shows the concept of coherence. Temporal coherence can be quantified through two important measures such as coherence length and time. This property can be improved by run the laser in single longitudinal and transverse mode.

3.2 Monochromatic

It is the most important property of laser and it can be measured by spectral line width. When the range of emitted frequencies is small by a light source, it is said to be high monochromatic. Laser beam normally have very few or single spectral lines with highly narrow widths as shown in **Figure 6**. Monochromaticity is most important because wide range of applications depends on this property such as interferometry, velocimetry, holography, separation of isotope and communications which require laser beam content. But this property is a not decisive factor for machining.

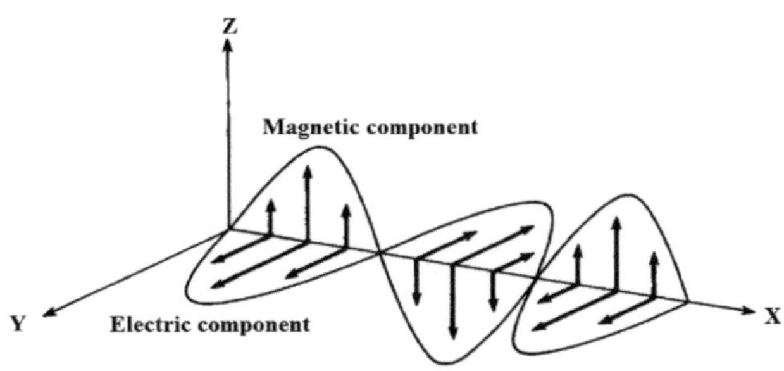

Figure 4.
Components of electromagnetic wave.

3.3 Low-diffraction or collimation

Directionality is a property by which a light beam bends after passing sharp corners of objects. Diffraction or scattering of light at sharp edges increase the distance

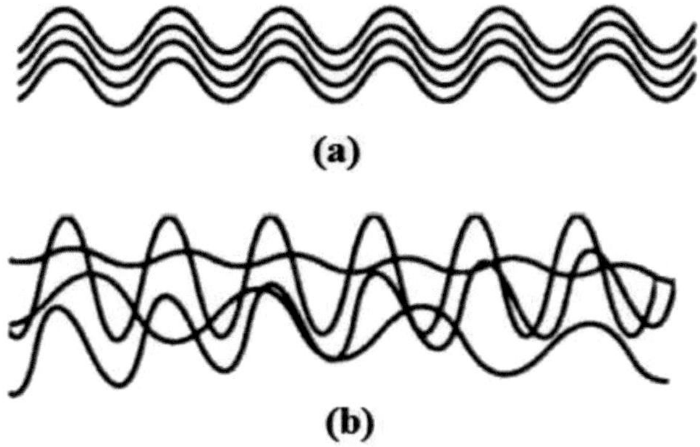

Figure 5.
Schematic of spatially and temporally (a) coherent light and (b) incoherent light.

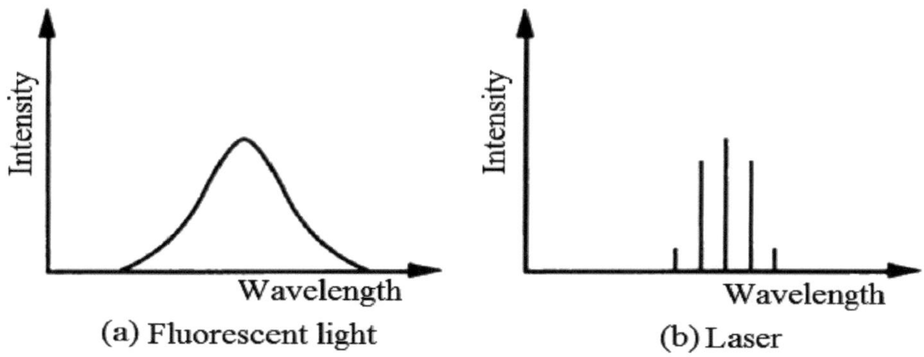

Figure 6.
Monochromaticity.

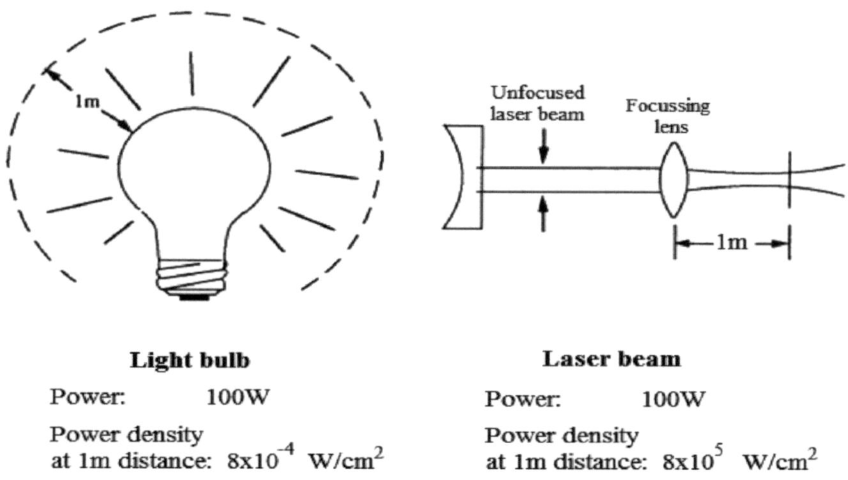

Figure 7.
Comparison of radiation from normal light bulb and a laser beam.

from light source therefore certain amount of energy is lost. But laser beams possess very low-diffraction property hence higher energy transfer can be effectively achieved. This directional characteristic is useful when directing the laser beam for machining applications.

3.4 Intensive radiance

The intensive radiance of a light is defined as the amount of power emitted per unit area for a given solid angle. The unit for radiance is watts per square meter per steradian. The angle by which a light beam is focused as a cone is called a solid angle. Since the intensity of photons is high in laser beam, it can have high output powers. Laser light source possess extreme amount of intensive radiance and transmitted through a small solid edge angle. This property makes it very convenient to be used for machining operations. **Figure 7** gives the comparison of power density transmitted by normal light source and a laser [3].

4. Types of laser

Lasers are classified based on the state of lasing medium used and the temporal mode. Based on the physical nature lasers are classified into solid-state lasers, gas lasers, semiconductor, and liquid dye lasers [4]. Based on temporal mode, further laser is categorized into two modes namely continuous wave (CW) and pulsed mode. Continuous mode emits the laser beam continuously without interruption whereas pulsed mode emits the laser beam periodically. **Tables 1** and **2** shows the important laser types along with their wavelengths.

In solid-state layers, the lasing medium is doped with very small number of impurity ions. Maimam has developed the first solid-state laser during 1960 which was a ruby laser. There are a number of laser types developed in the solid-state category in which Nd:YAG is majorly used for LBM applications. Solid-state lasers such as Nd:YAG, ruby and Nd-glass are highly used for machining metallic materials. Nd:YAG lasers can also be used to ceramic materials. Gas lasers are grouped

Solid-sate lasers.		Gas lasers	
Lasing medium	**Wavelength (nm)**	**Lasing medium**	**Wavelength (nm)**
Ruby	694	ArF	191
Alexandrite	700–820	KrF	249
Ti-sapphire	700–1100	XeCl	308
Nd-YLF	1047	XeF	351
Nd:YAG	1064	Argon	488, 514.5
Nd:glass	1062	Krypton	520–676
Er-YAG	2940	HeCd	441.5, 325
—	—	Copper vapor	510.6, 578.2
		Gold vapor	628
		HeNe	632.8
		CO_2	10,600

Table 1.
Solid-sate and gas lasers with their wavelengths.

Semiconductor lasers		Liquid dye lasers	
Lasing medium	**Wavelength (nm)**	**Lasing medium**	**Wavelength (nm)**
AlGaInP	630–680	Stilbene	403–428
AlGaAs	780–880	Coumarin 102	460–515
InGaAs	980	Rhodamine 6G	570–640
InGaAsP	1150–1650	—	—

Table 2.
Semiconductor and liquid dye lasers with their wavelengths.

into three categories based on the composition such as neutral atom, ion, and molecular. Gas lasers generally can be of CW or pulsed mode lasers and available with axial flow, transverse flow and folded axial flow in construction. CO_2 laser is the most commonly used gas laser for machining plastics, ceramics, nonmetals and sometimes organic materials also.

Semiconductor lasers, though made of solid materials the working principle are different from solid-state lasers. It is based on radiative recombination of charge carriers. Unique characteristic of a semiconductor laser is that they are capable of producing wide beam divergence angles around 40°. Comparing to other types of lasers, liquid-state lasers are easier to fabricate. Main advantages of liquid-state lasers are ease cooling and replenishment in laser cavities. Spectral properties of liquid organic molecules enable liquid dye lasers to get tuned within wide range of wavelengths from 200 nm to 1000 nm. The detailed working principles of these lasers are beyond the scope of this chapter and can be found in any standards texts.

5. Material removal using laser

5.1 Construction of LBM

Laser beam machining is a nonconventional, advanced machining process wherein there are essential parts required to construct a complete LBM setup. A pumping medium or lasing medium that contains large quantity of atoms is a primary component to produce laser light. For exciting the atoms in lasing medium, a flash lamp or flash tube is needed and it should be connected to the controlled high voltage power supply. Based on the type of operating mode (either pulsed mode or CW) a capacitor can be integrated to the power circuit. A typical solid-state LBM setup is schematically shown in **Figure 8**.

5.2 Mechanism of material removal in LBM

Laser based machining processes is identified as a material removal technique in industrial application. Materials removal is accomplished by the interaction between the laser beam and work material. It is severely a localized thermal process. Higher amount of light energy is received by base material and then higher heat is created between the locality of interaction while hitting the laser source on the base material. Due to highly elevated temperature at the beam spot, the material becomes soft, melt, burn and vaporized. Additionally, the interaction of laser beams and work material is associated with the material removal by photochemical process which is often called photo ablation. **Figure 9** schematically represents the effects of laser beam-work material interaction [5].

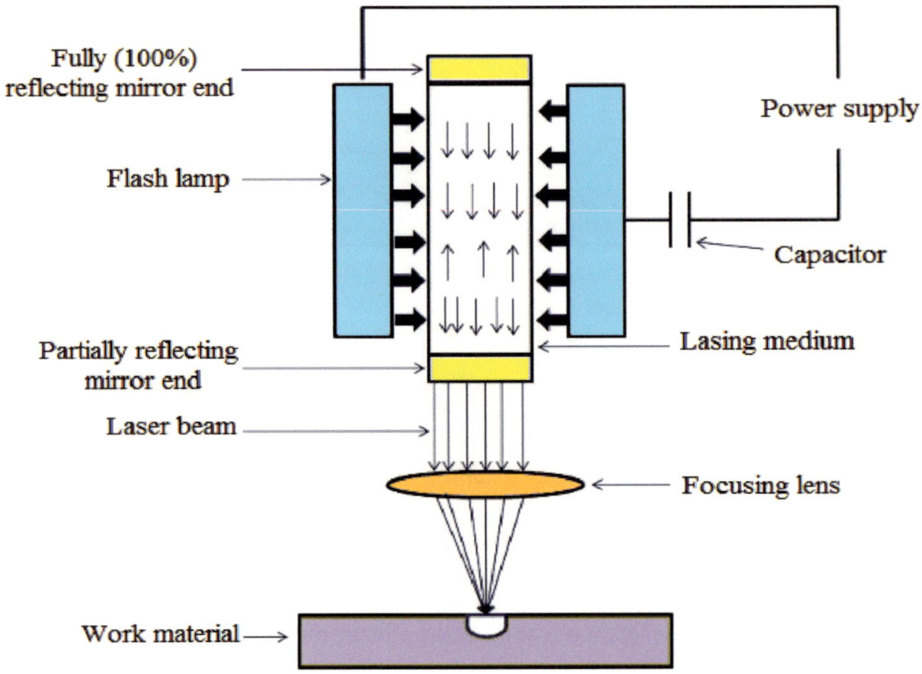

Figure 8.
Laser beam machining setup.

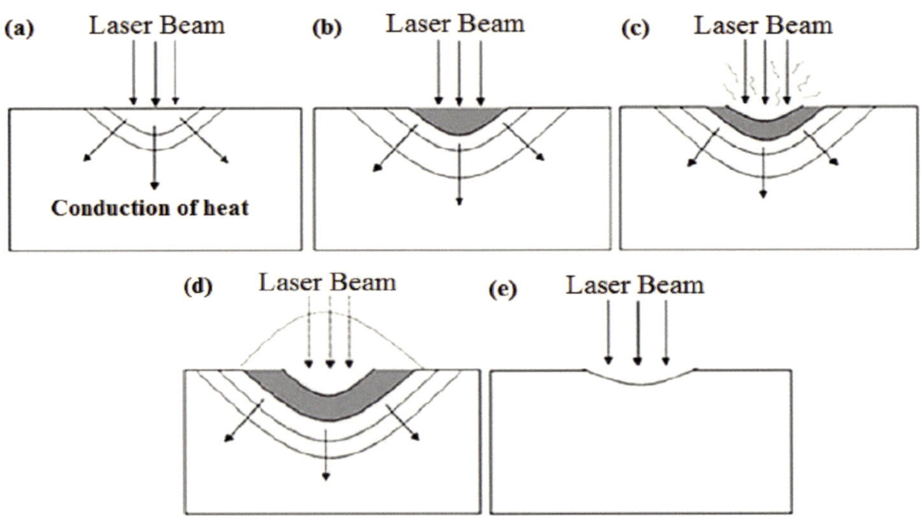

Figure 9.
Laser beam-work material interactions: (a) heating, (b) melting of surface, (c) vaporization of surface, (d) formation of plasma, and (e) ablation.

The parameters of LBM such as intensity of laser light, distribution of beam, spot size, scanning speed, and relative motion between laser beam and work piece can be changed according to the requirements for different work materials. As presented in the introduction section, lasers are replacing conventional machining processes due to many advantages. Many developments have been made in the laser technology to shorten the pulse time for different machining processes. Longer pulse duration increases the heat affected zone (HAZ) and leaves high thermal

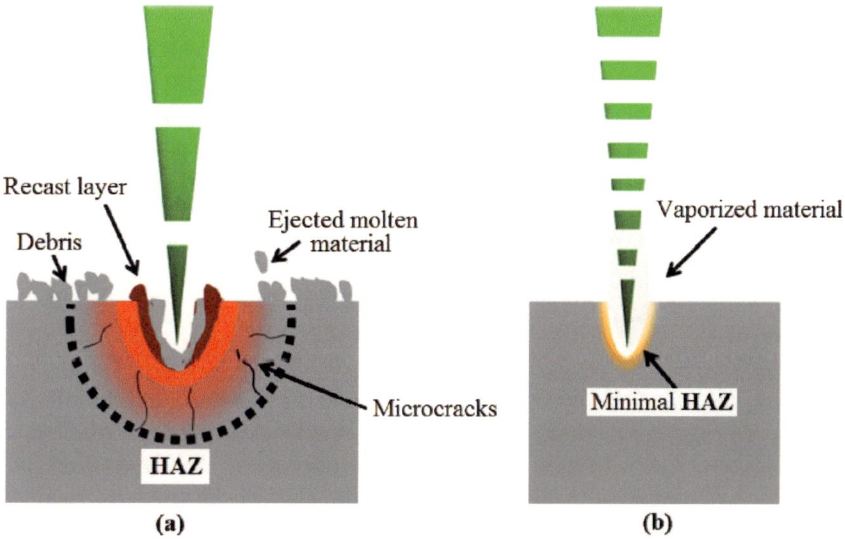

Figure 10.
Difference between the effects of (a) long-pulsed and (b) short-pulsed lasers.

stresses resulting in crack and void formation, and surface debris. Short pulse duration leads to lesser thermal conduction thus resulting in precise machining operation and good surface finish. **Figure 10** shows the difference between the effects of long and short pulse durations [6].

5.3 Types of LBM techniques

Machining using laser is generally categorized into three types namely one-dimensional, two-dimensional and three-dimensional machining processes. In one dimensional machining process, the laser beam will have no relative motion with the work piece material. In this relatively stationary arrangement, the erosion front is located at the work piece and focused laser beam removes the material in the path it propagates through which is a straight line. Hence one-dimensional LBM process is generally used for drilling applications. In contrast, the work piece also will move along with laser source in two-dimensional LBM process. The erosion front placed on the beam edge and the material removal happens in a two-dimensional plane

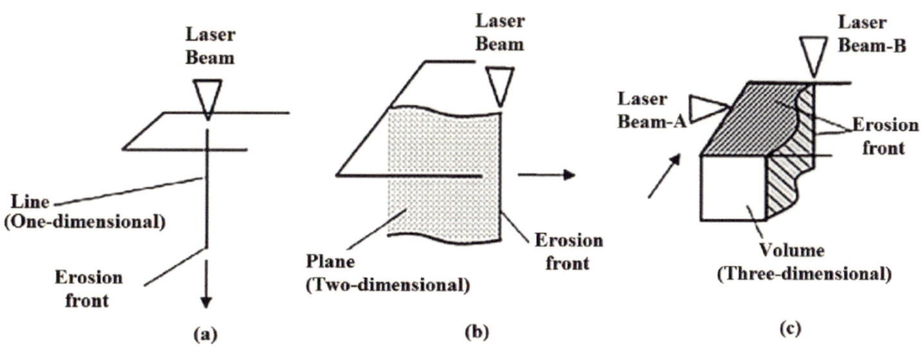

Figure 11.
Schematic of (a) one-dimensional (drilling), (b) two-dimensional (cutting), and (c) three-dimensional (milling) machining operations.

resulting in a creation of two-dimensional surface as shown in **Figure 11**. Two-dimensional LBM is most suitable for cutting operations. Three-dimensional LBM uses two or more sources of laser beams. Each laser beam forms two-dimensional surfaces according to their relative motion with the work piece. When the surfaces formed by each laser beams intersects a three-dimensional space is created that defines the shape of material to be removed. Three-dimensional LBM process is generally used for milling process. For better understanding of different types of LBM processes a schematic representation is given in **Figure 11** [7].

6. Applications

In general, the use of lasers found in many applications includes chemical, biochemical, optics, medical, military operations, polymer sciences, nuclear physics [8–12] and so on. In manufacturing, lasers are successfully applied for material removal, metal joining, cladding and alloying processes. Specifically, this chapter discusses the material removal applications of lasers. Drilling, grooving, cutting, three-dimensional machining operations such as lathe and milling operations, micro machining and laser assisted machining processes are the extended applications of lasers in LBM [3].

6.1 Drilling using LBM

One of the major advantages in drilling using lasers compared to conventional machining process is the aspect ratio (max 1:20) and small size of the hole drilled. Both continuous and pulse laser are used for drilling operations in which pulsed laser gives lesser plasma generation. The types of drilling operations that can be performed using LBM are single-pulsed drilling, percussion drilling, trepanning and helical drilling. When laser beam is focused into the material, the temperature is created by absorbing the photons. The material melts and vaporizes when the temperature exceeds the melting temperature of the material. If the radiation of laser is set lesser than particular threshold (106 W/cm^2 for steels), the material melts but not vaporizes and using a jet of gas the molten material is ejected [13]. The single pulse drilling process makes either through or blind holes with less than 1:15 aspect ratio. This is a rapid drilling process mainly suitable where production rate is more important than quality. Single pulse drilling is mostly adapted in automotive industry for processing connecting rods and filters. Percussion drilling uses pulsed-lasers' focus on the same spot to produce a hole while maintaining a balance between throughput and quality. Due to major advantages like its precision and quick processing capability, percussion drilling is adapted in making holes in the blades of turbine-airfoil. Though it has major advantages, there are drawbacks reported such as dross, spatter, and tapering.

Trepanning technique is another hole making technique where the material removal is performed on the circumference of any circle to make holes of higher diameters. It is considered to be a standard technique for making holes around 500 micrometers diameter. The nanosecond pulsed laser source is utilized for material removal around the circumference hence the drawbacks of percussion drilling remains in this application also. Trepanning technique reduces the taper effect and produces more jagged edge quality. To overcome the drawbacks of trepanning a relatively new technique called helical drilling is introduced. Helical drilling follows multitude ablation steps. The advantages of helical drilling over trepanning using percussion drilling are improved circularity of drilled holes, minimized loads on opposite walls and more importantly reduced recast layers or sometimes completely

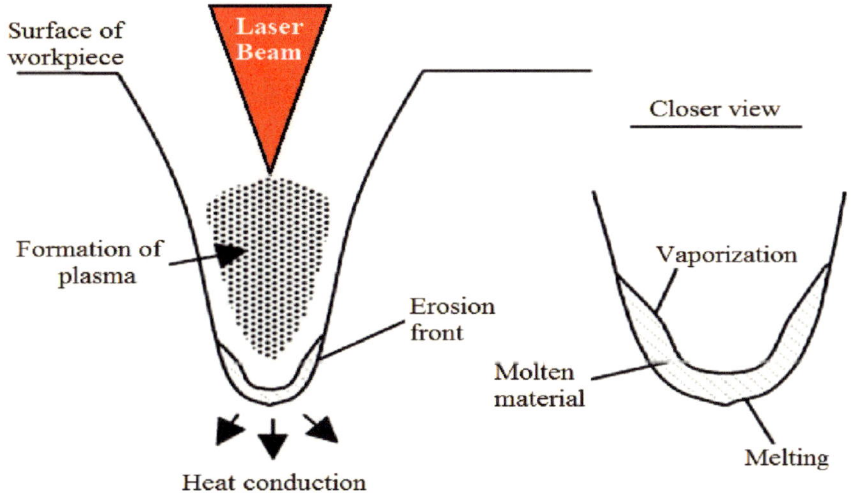

Figure 12.
Schematic of laser drilling.

avoided. Helical drilling is more preferred in the case of laser beam diameter is very near to helical diameter at focus point. Energy balance is important in laser drilling among the energy released by laser beam, energy absorbed by material, energy lost to the surrounding and the energy utilized for melting (phase changing) the material as shown in **Figure 12**.

6.2 Cutting using LBM

Cutting is an essential operation in any material removal processes. A relative motion between work piece and the laser beam is required to produce a two-dimensional working plane where the material removal takes place. During relative motion of laser beam and work piece, a kerf is produced which removes the material in its path. Complex two-dimensional shapes can be cut from the flat work piece

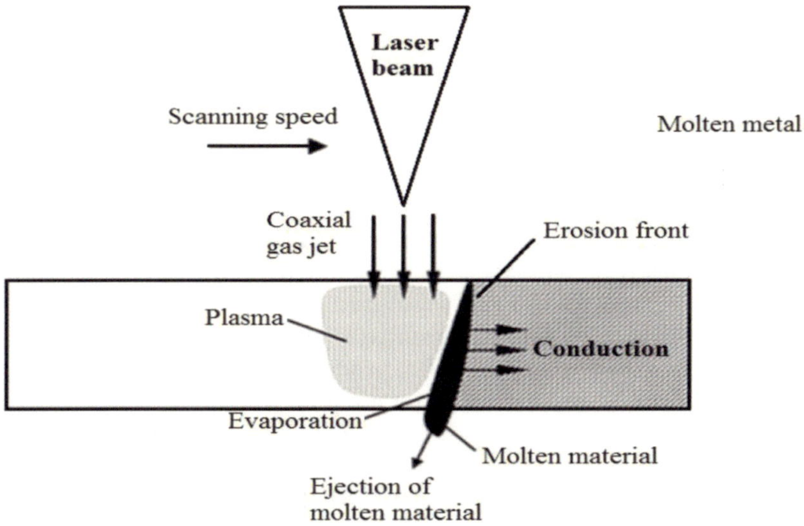

Figure 13.
LBM for cutting operation.

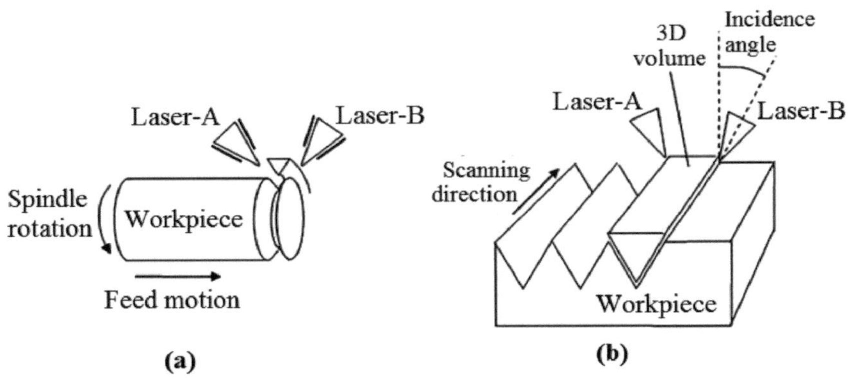

Figure 14.
Three-dimensional LBM for (a) lathe operation and (b) milling operation.

materials where mechanism of material removal is similar to drilling operation. In contrast to the drilling process, the erosion front is located at the front of line of laser beam as shown in **Figure 13**. However, the temperature field and erosion front is fixed based on the coordinate moves along with produced laser source that ensures steady state process. The erosion front molten material is flushed away using a gas jet during the cutting process.

6.3 Three-dimensional LBM

Three-dimensional LBM uses two or more number of laser beams simultaneously focused to obtain an intersected volume for material removal. To precisely create such volumes with relative motions, highly accurate optical manipulating systems are therefore necessary. Recent systems equipped with optical scanning systems have high level of control over the motion of laser beams which enables efficient and effective machining operations. The material removal using these tools is referred as 3-dimensional (3D) laser material processing. In general, the 3D laser material processing is grouped into various categories such as laser beams along with 3D LBM, 5-axis heads along with 3D processing workstation and 3D remote laser processing. **Figure 14** illustrate the graphical picture of two-beam laser machining processes for lathe and milling operations. Each beam creates a groove like volume of material removal when they intersect with some incidence angle. The incidence angle may be changed and dynamically varied along with relative motion to get intricate shapes of machining.

LBM is successfully adapted in micromachining field due to its high flexibility to automation and high degree of radiance. Laser beam micromachining is capable of producing parts with sizes ranging from micro to sub-micro scales. It usually employs the pulsed lasers with an average power of less than 1 kW. The pulses of femtosecond duration are widely used for micromachining. Micromachining can be performed on wide range of materials such as metals, glasses, diamond and other difficult to machining materials. Laser-assisted manufacturing (LAM) is another technique helps to enhance the maximum productivity, quality with minimized machine tool vibrations, machining forces and tool wear. LAM is also an effective technique to machine brittle materials without cracks and failure. This hybrid machining process, laser beam is focused on the work piece just before the cutting tool engages. Scanning of laser initially heats up the work therefore helps in plastic deformation rather than brittle deformation during machining. The LAM processes are suitable for brittle and hard type of material such as ceramics, nickel alloys and the higher amount of silicon element material.

7. Advantages

LBM is an excellent manufacturing technique to process wide ranges of difficult to machine materials especially ceramics and advanced composite materials. LBM technique is capable of machining intricate shapes that cannot be reached or processed by conventional machining processes. LBM is an alternate to conventional machining processes due to many advantages as follow:

- Due to precise machining capability, LBM can produce excellent surface finish therefore post processing can be eliminated.

- LBM is a clean manufacturing technique due to less environment pollution and no requirement of chemicals or solvents for machining.

- LBM can be easily automated for higher productivity and to achieve high speed machining.

- It uses no cutting tool therefore no cutting forces involved during machining. This phenomenon helps to avoid heavy construction of machine tools, physical damages, vibrations, frequent tooling requirements.

- Degree of accuracy in machining complex geometry is high.

- LBM depends on thermal and few optical properties of work material rather than mechanical properties such as hardness and brittleness. As a result, most of the materials with any degree of mechanical properties with lower diffusivity and conductivity can be machined.

- Wide range of materials from plastics to diamond can be machined.

- Residual stresses caused due to HAZ are very less in LBM.

- Machining micro features with large aspect ratio is possible with LBM.

8. Limitations

There are many issues and limitations associated with the aforementioned LBM technique. The major issues are produced accuracy, achieved surface quality and rate of material removal. The erosion front is the main factor decides the amount of material removal in LBM technique. In one-dimensional machining, the speed of propagation in erosion front in the straight line decides the rate of material removal. In another hand, the scanning speed plays a significant role in metal removal during the two-dimensional machining processes. Similarly, the laser scanning speed is produced the intersecting surfaces for volume formation and the decisive factor for material removal rate during the 3D machining processes. Controlling the LBM parameters for a balanced and effective machining is a real challenge faced by industries. Secondly, the dimensional accuracy is affected by the kerf shape of laser which leads to tapered holes instead of narrow holes. Surface quality is the other important aspect of machining, which is measured by surface roughness, formation of dross and the HAZ. Since LBM is completely thermal based machining process it also has several limitations such as,

- Minimum amount of metal removal
- Investment cost is high
- Highly skilled operator is required
- Maintenance cost is high
- Power consumption for LBM is high
- Transparent and greatly reflective materials cannot be machined using LBM
- Applications related to machining thicker materials are very limited

9. Conclusions

The chapter presented an overview of the LBM technique and the principle of laser production, properties, types of lasers, and its application in machining field. The advantages and limitations are also discussed at the end. Based on the discussions from the presented sections, the following conclusions are made regarding LBM.

- LBM is an effective technique for processing complex geometries of different materials with superior properties. While this technique is mostly advantageous in the field of machining, it also possesses few disadvantages such as low energy efficiency, low material removal rate which affects productivity, quality concern due to diverged or converged laser beam.

- LBM depends upon many important parameters such as wavelengths of lasers used, scanning speed, type of laser beam (pulsed or CW), pulse duration, assist gas and its flow, material properties and physical dimensions to assess some of the performance characteristics like surface quality, thermal stresses due to HAZ, formation of dross and defects.

- Excellent flexibility to automation of LBM enables it to be used in advanced machining applications like micromachining with superior level of accuracy.

Acknowledgements

The authors wish to thank the Ministry of Science and Technology, Taiwan ROC for the financial support to carry out this work.

Conflict of interest

The authors declare that they have no conflicts of interest in the work.

DOI: http://dx.doi.org/10.5772/intechopen.93779

Author details

Natarajan Jeyaprakash[1*], Che-Hua Yang[1,2] and Manickam Bhuvanesh Kumar[3]

1 Additive Manufacturing Center for Mass Customization Production, National Taipei University of Technology, Taipei, Taiwan, ROC

2 Institute of Manufacturing Technology, National Taipei University of Technology, Taipei, Taiwan, ROC

3 Department of Production Engineering, National Institute of Technology Tiruchirappalli, India

*Address all correspondence to: prakash84gct@gmail.com; prakash@ntut.edu.tw

References

[1] Nagimova A, Perveen A. A review on laser machining of hard to cut materials. Materials Today: Proceedings. 2019;**18**:2440-2447

[2] Kuar A, Doloi B, Bhattacharyya B. Modelling and analysis of pulsed Nd: YAG laser machining characteristics during micro-drilling of zirconia (ZrO_2). International Journal of Machine Tools and Manufacture. 2006;**46**(12-13):1301-1310

[3] Chryssolouris G. Laser Machining: Theory and Practice. New York, United States: Springer-Verlag; 2013

[4] Dahotre NB, Harimkar S. Laser Fabrication and Machining of Materials. United States: Springer Science & Business Media; 2008

[5] Chryssolouris G, Stavropoulos P, Salonitis K. Process of laser machining. In: Nee A, editor. Handbook of Manufacturing Engineering and Technology. London: Springer London; 2013. pp. 1-25

[6] Agrawal R, Wang C. Laser Beam Machining. In: Encyclopedia of Nanotechnology. 2nd ed. The Netherlands: Springer; 2016. pp. 1739-1753

[7] Chryssolouris G, Anastasia N, Sheng P. Three-dimensional laser machining for flexible manufacturing. In: Conference on Laser Application in the Automotive Industries. Aachen: SAE International; 1991. pp. 247-253

[8] Chow WW, Jahnke F. On the physics of semiconductor quantum dots for applications in lasers and quantum optics. Progress in Quantum Electronics. 2013;**37**(3):109-184

[9] Kaushal H, Kaddoum G. Applications of lasers for tactical military operations. IEEE Access. 2017;**5**:20736-20753

[10] Moore CB. Chemical and Biochemical Applications of Lasers V1. Vol. 1. United States: Elsevier; 2012

[11] Murnick DE, Feld MS. Applications of lasers to nuclear physics. Annual Review of Nuclear and Particle Science. 1979;**29**(1):411-452

[12] Vij D, Mahesh K. Medical Applications of Lasers. United States: Springer Science & Business Media; 2013

[13] Ready J. LIA Handbook of Laser Materials Processing. Berlin, Heidelberg, German: Springer-Verlag; 2001

Chapter 6

Nanoparticle Formation and Deposition by Pulsed Laser Ablation

Toshio Takiya and Naoaki Fukuda

Abstract

Pulsed Laser Ablation (PLA) in background gas is a good technique to acquire specific nanoparticles under strong non-equilibrium states. Here, after a history of PLA is mentioned, the application of nanoparticles and its deposition films to the several fields will be described. On the target surface heated with PLA, a Knudsen layer is formed around the adjacent region of the surface, and high-pressure and high-temperature vapor atoms are generated. The plume formed by evaporated atoms blasts off with very high-speed and expands rapidly with a shock wave. A supercooling phenomenon occurs during this process, and number of nucleus of nanoparticle forms in vapor-phase. The nuclei grow by the condensation of vapor atoms and deposit on a substrate as nanoparticle film. If the radius of nanoparticle is uniformized, a self-ordering formation can be shown as a result of interactive process between each nanoparticle of the same size on the substrate. In this chapter, the related technology to realize a series of these processes will be expounded.

Keywords: PLA, PLD, nanoparticle, deposition, non-equilibrium, evaporation, laser plume, shock wave

1. Introduction

Pulsed Laser Deposition (PLD), which is a film-forming technique by using PLA, has been experimented since the 1960s after the invention of the laser oscillator. Afterwards, in the wake of the realization of Q-switch Nd: YAG laser in 1970s and the electric discharge pumping high-speed repeatable excimer laser in 1980s, PLD has become a powerful tool to fabricate high-performance films such as semiconductors, heterostructure and superlattices. The success of high-temperature superconducting films in the mid-1980s led to the flourishing of laser ablation research.

In the early 1990s, the research on PLD application has made several developments such as heteroepitaxial films, perovskite oxide films, nitride films and diamond-like carbon films. Although the research on nanoparticles derived from the laser ablation had already been conducted as part of these developments, the discovery of Buckminster fullerene [1] was undoubtedly an important milestone in the following development of nanoparticle applications. In the 21st century, effect which shock waves have on the formation way of nanoparticle during laser ablation has been investigated [2, 3]. Several researches [4–6] in which the shock waves are positively used to form monodispersed nanoparticles and composite nanoparticles like core-shell type have been conducted. The laser ablation in nanoparticle research is regarded as the promising method at present which is possible to fabricate a novel function device in the area of electricity, semiconductor, energy and so on.

We have known PLD as a process during which a solid target is vaporized with laser irradiation, and then the nanoparticle formation in the gas phase is occurred, followed by its soft-landing on a substrate. But each process remains unclear to us. In order to apply the laser ablation method for the fabrication of functional materials, it is important to understand each step of the process so that we can put them to practical use. In this chapter, we divide the formation process of nanoparticle films as following steps for further studies: (1) temperature increase of solid surface by laser irradiation, (2) evaporation of the surface and its conversion to kinetic energy, (3) plume expansion with shock wave propagation, (4) supercooling of evaporated gas, (5) uniform sized nanoparticle formation, and (6) nanoparticle deposition and self-ordering on the substrate.

2. The formation process of nanoparticles by laser ablation

2.1 Laser irradiation analysis

The process of laser ablation commences from the interaction of lights and solids. When a laser irradiates on the surface of a solid target, the electromagnetic energy of the laser beam firstly turns into the excitation energy of electron, although the process may vary depending on the thermodynamic state of the solid. The energy conversion in the solid from the electronic excitation to the lattice vibration completes within some picoseconds. If the laser has nanosecond pulse, therefore, it is not necessary to take into consideration the non-equilibrium of the laser irradiation. Then, when viewed as a one-dimensional problem in the depth direction, the solid temperature can be described as the usual one-dimensional unsteady-state heat conduction Equation [7].

$$\frac{\partial T}{\partial t}=\frac{\partial}{\partial x}\left(\kappa\frac{\partial T}{\partial x}\right)+\frac{\alpha}{c_p\rho}I(x,t) \tag{1}$$

in the above equation, T stands for temperature, while κ for thermal conductivity, α for optical absorption coefficient, c_p for the specific heat, ρ for the density of the target material, and I for the value of laser energy, according to Lambert's law, along the depth direction after deduction of the loss caused by surface reflection.

On the other hand, a boundary condition is applied on the surface of target, where quantity of heat is removed from the surface corresponding to the heat of melting and evaporation. In addition, since the physical properties such as thermal conductivity are generally different between the solid and liquid phases, the equation should be applied separately in each phase [8].

As will be discussed in the next section, since the evaporating atoms during laser ablation are highly directional, the one-directional analysis is effective in most cases. However, in a certain case, one-dimensional analysis fails to take into consideration some factors, such as the thermal conductivity in radial direction which becomes dominant factor when the laser irradiation time becomes longer. Formerly, Houle and Hinsberg [9] applies a two-dimensional axisymmetric model with probabilistic algorithms to the analysis of laser ablation. This allows phenomena with different characteristic time such as absorption, melting, evaporation, and thermal conduction to be analyzed at the same time marching [10].

2.2 Knudsen layer

If the vapor pressure on the surface of the solids during evaporation is equal to the ambient pressure, the Maxwell velocity distribution for both evaporating and recondensing gas takes half shape of the distribution f^+ or f^-, being opposite in sign to each other, and the average value of velocity for these gases ought to be zero. In other words, if we consider the gases evaporating from the target surface and recondensing into it as a mixed gas, it is in an equilibrium state in terms of translational motion. However, when evaporation occurs drastically, as in the case of laser ablation, the balance of velocity distribution breaks down and a one-sided distribution appears for the evaporating gas, while the recondensing gas becomes remarkably little [11] so as to take the following velocity distribution [12].

$$f_s^+ \propto \exp\left[-\frac{2E_I + m\left(v_x^2 + v_y^2 + v_z^2\right)}{2kT_s}\right]; \qquad (2)$$

$$v_x \gg 0, -\infty < v_y, v_z < \infty,$$

where direction x is normal to the target surface, directions y and z are in parallel to the surface, as an index of kinetic velocity v. While E_I is the accessible internal energy, T_s is the surface temperature, m is the atomic mass and k is Boltzmann constant. The movement of the evaporating atoms in vacuum and atmosphere is shown in **Figure 1** [13], by way of comparison. In vacuum as **Figure 1(a)**, the flow velocity as a macroscopic fluid is zero, while the kinetic velocity of individual atoms which are without interferences from the other atoms to form a whole gas flow (free molecular flow) is very high. When the gas density becomes high, as shown in **Figure 1(b)**, the velocity vector of atoms which keeps positive direction near the surface, gradually changes to negative direction with increasing distance from the surface due to the presence of recondesing atoms.

Meanwhile, the atomic velocity transits to the fluid velocity and the overall gas flow velocity increases. Thus, the velocity distribution of the evaporating atoms is relaxed in a very thin layer near the solid surface and finally reaches a shifted Maxwell equilibrium distribution as following [12]:

$$f_K^{\pm} \propto \exp\left[-\frac{2E_I + m\left\{\left(v_x - u_K\right)^2 + v_y^2 + v_z^2\right\}}{2kT_K}\right]; \qquad (3)$$

$$-\infty < v_x, v_y, v_z < \infty,$$

where u is flow velocity and index K represents the state value at the very thin layer, which has a thickness corresponding to several mean free path and is called the Knudsen layer.

At the boundary of the Knudsen layer, the velocity of the gas u_K is equal to the speed of sound a_K, and the subsequent flow state will be determined by the temperature of the target surface, the type of ambient gas and its pressure [12, 14]. By solving the Boltzmann equation for the Knudsen layer, the velocity and angular directional distributions of evaporating gas near the solid surface have also been analyzed [15, 16].

2.3 Plume expansion and shock wave propagation

The configuration of plume (vapor atomic mass) formed by the vapor atoms emitted from the laser-irradiated area and shock wave formed by the piston effect of the plume is shown in **Figure 2** [17]. From the laser irradiation zone shown in the leftmost part of the figure, since the vapor atoms emerge from the Knudsen layer in a certain angular distribution of velocity, the plume spreads out to the lateral

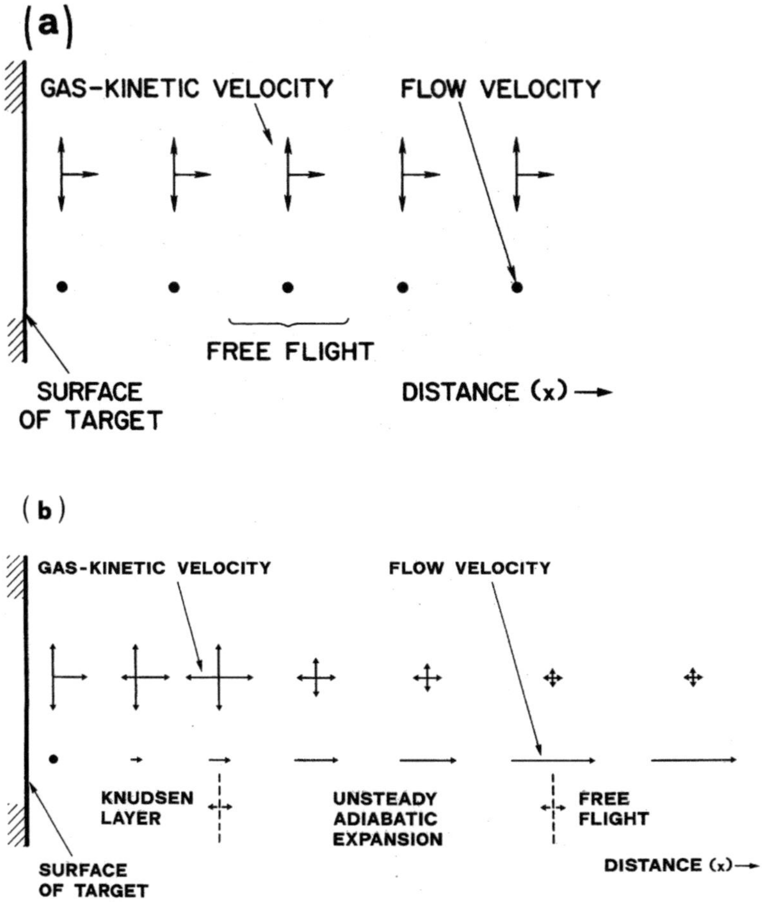

Figure 1.
(a) Schematic of representation of vapor atoms emitted from a target surface which enter immediately into free flight, in the case of evaporation in vacuum. (b) Schematic representation of Knudsen layer formation followed by an unsteady adiabatic expansion and free fight, in the case of evaporation in high density regime [13].

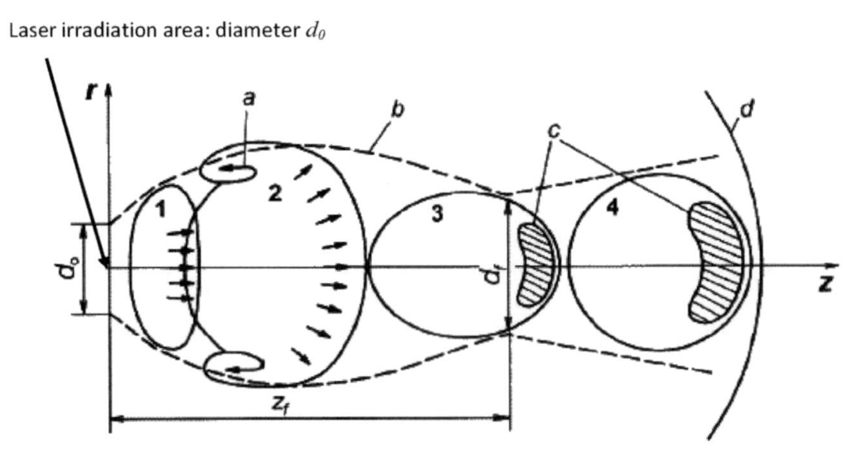

Figure 2.
Schematic drawing of laser plume expansion from the viewpoint of high-speed hydrodynamics [17].

direction along the central axis. In general, under the conditions of laser ablation, the pressure of the plume developed from the target is extremely higher than the ambient pressure, which is the so-called under-expansion jet in term of high-speed fluid dynamics (region 1 in **Figure 2**). Thereafter, the plume continues to expand, then it overinflates and forms a mushroom-like vortex (region 2 in **Figure 2**). The pressure of the overinflating plume becomes negative in comparison to the surrounding gas, which causes the plume to contract, resulting in a minimum diameter of the plume (region 3 in **Figure 2**). After this, the plume expands again and gradually decays (region 4 in **Figure 2**). At this stage, the resistance of the ambient gas is strong for the progress of the plume. The ambient gas atoms diffuse into the plume's front edge (hatched area in **Figure 2**) and form a diffusion region there. Since radicals of vapor atoms are generated in this diffusion region and plunge in collisional relaxations with the ambient gas atoms, luminescence can be likely observed in the region.

Since the atoms fall on a substrate from the luminescence zone, photogenic property in this zone would have a powerful effect on the characteristics of the film produced by laser ablation. The phenomenon has been analyzed through direct imaging by Intensified Charge Coupled Device (ICCD) and Laser Induced Fluorescence (LIF) [18–23]. The formation of nanoparticles is also thought to occur in this region [24].

2.4 Formation of nanoparticles

As described in the previous section, the formation process of the nanoparticle is highly related to the thermodynamic state in the dilution zone of the vapor atoms and ambient gas atoms. Hagena and Obert [25] conducted experiments using a supersonic nozzle to summarize the relationship between the size of atomic nanoparticles and the thermodynamic state as similitude rules in which parameters of molecular movement such as an interatomic potential are primary variables. Thus, estimating the equilibrium concentration of a nanoparticle from the thermodynamic conditions of the surrounding environment is useful for practical purposes and has been discussed in many publications. The rate constant for the nanoparticle equilibrium concentration is essentially based on the assumptions of statistical mechanics and is expressed as follows [26].

$$K_g = \frac{Q_t Q_r Q_v}{(q_t)^g} \exp\left(-\frac{E_g}{kT}\right) \tag{4}$$

Here, K_g is the rate constant of the equilibrium concentration in a nanoparticle consisted of atoms of the number g, while Q is the partition function of the nanoparticle with the subscripts t, r and v representing the translation, rotation and oscillation respectively. The variable q stands for the partition function of a mono-atom, E_g for the formation energy of the nanoparticle, k for the Boltzmann's constant, and T for the temperature of the system. The above equation describes the process in which the translational kinetic energy of a mono-atom transformed into the internal energy of the nanoparticle when it is captured by the nanoparticle. But there does not seem to be any detailed descriptions of this process in the classical theory of nanoparticle formation.

On the other hand, Eq. (4) refers to the nanoparticle concentration based on the steady-state theory. Generally speaking, the length of time it takes to form nanoparticle is not considered in this equation. However, for high-speed phenomena such as laser ablation, it is necessary to estimate the nanoparticle formation time to determine if the balance between condensation and evaporation on the surface of nanoparticle is kept or not. If we are assuming that the velocity of vapor atoms is represented in the equilibrium Maxwell distribution, we can estimate the approximate nanoparticle formation time τ with the following equation.

$$\tau = \frac{\rho_c r\sqrt{2\pi kT}}{p\sqrt{m}} \tag{5}$$

In this equation, ρ_c stands for the internal density of the nanoparticle, while r the radius of the nanoparticle, p the vapor pressure, and m the mass of the vapor atoms. In order to obtain an exact solution, Gillespie [27] analyzed the process of nanoparticle formation as a random walk problem. However, no matter how rigorous the probabilistic analysis is, it is still based on classical nucleation theory. To address the problem related to the formation of nanometer-sized particle in non-equilibrium, the following issues should be considered.

a. The nucleation rate equation based on steady-state theory is valid only when the formation time of a critical nucleus is sufficiently short compared to the representative time of the system.

b. The classical theory is an available model at the range of relatively low supersaturation. In the case of high supersaturation, the internal degrees of freedom of the nanoparticle must be taken into an account, because differences are created among translational, rotational and vibrational temperature.

c. In spite of small systems, macroscopic physical properties and coefficients such as surface free energy, evaporation enthalpy, and condensation coefficient besides macroscopic concepts such as the Thomson–Gibbs formula are used. We have to correct these values and formula in conformity with the extent of non-equilibrium.

In recent years, with the development of computers, molecular dynamics or Direct Simulation Monte Carlo analyses which take into consideration the internal degrees of freedom of nanoparticles have been made [28], through which the Gibbs free energy and nanoparticle concentrations are elucidated under more realistic conditions.

3. Films fabricated by monodispersed nanoparticle beams

3.1 Monodispersed nanoparticle by size classification

In general, the size distribution of nanoparticles $f_x(x)$ growing in vapor phase is known to be described as the log-geometric distribution as follow [29].

$$f_X(x) = \frac{1}{x\sqrt{2\pi}\ln\sigma_g}\exp\left[-\frac{1}{2}\left(\frac{\ln(x/m_g)}{\ln\sigma_g}\right)^2\right] \quad x \geq 0 \tag{6}$$

Here m_g is the geometric mean and σ_g is the geometric standard deviation. A model has been proposed to explain the log-geometric distribution by assuming that the growth formula of a nanoparticle follows a certain Equation [30, 31]. It has also been experimentally confirmed that the size distribution of nanoparticles generated by laser ablation is log-geometric [32].

While sizes of nanoparticles have a certain distribution, the physical property of the nanoparticle may vary dramatically depending on its size. Thus, the sizes of nanoparticles should be in uniform when we need deal with them as macroscopic materials like nanoparticle films. Hence, in the early 2000s, some kinds of device to make the nanoparticle size in uniform has been actively invented and designed.

In those times, the size selection of aerosol particles is generally done in an electric field by deflecting charged particles generated from a source of the outbreaks such as combustion gas in Diesel engine. Afterward, the Differential Mobility Analyzer (DMA), as shown in **Figure 3** [33], which has been used to measure aerosol particles in the atmosphere, has been improved to the desired extent for applying on nanoparticles [34–36]. It is able to measure the particle size distribution by using the natural law in which a flying charged nanoparticle would land on different locations according to the balance between the electric mobility and fluid resistance, being deflected by an electric field [37]. Camata et al. [38] used a DMA system to estimate the geometric standard deviation of silicon nanoparticles with an average particle diameter of 2.8 nm and got a number of the geometric standard deviation between 1.2 ~ 1.3. Suzuki et al. [39] used DMA system on silicon nanoparticles and obtained a result of an average particle diameter of 2.8 nm and a geometric standard deviation of 1.2. A consistent system from a source of nanoparticles to its sampler is shown in **Figure 4** [34]. The nanoparticles generated by laser ablation pass through the gas phase annealing system, where the particle morphologies are controlled, and move to DMA to be classified by its size, and finally reach the particle sampler and the measurement systems.

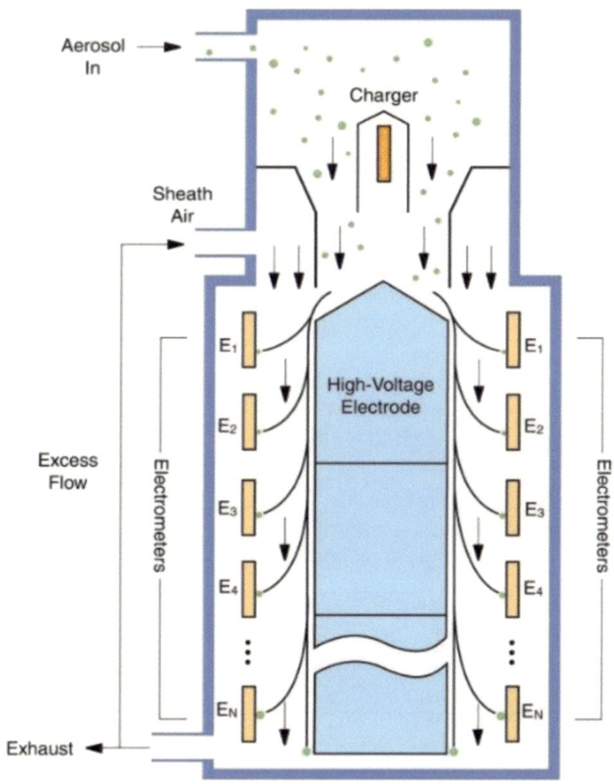

Figure 3.
Illustration of instrument capable of measuring size distributions of aerosol particles on basis of DMA [33].

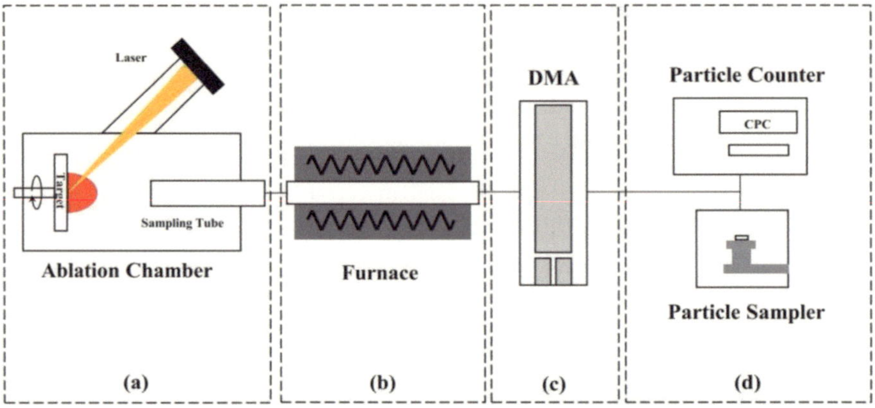

Figure 4.
Schematic of nanoparticle synthesis process using laser ablation, which is composed of (a) particle generator, (b) gas phase annealing, (c) particle classifying, and (d) particle measurement and correction [34].

Although the above methods require nanoparticles being charged in some way to select the size, the flank attack method proposed by Wu et al. [40] can be used for neutral nanoparticles. In this method, the nanoparticle beam is intersected with the atomic beams of inert gas such as argon or neon, and the nanoparticle can be sorted out in accordance with the size following the law in which the smaller the nanoparticle size, the greater the deflection angle of the beam.

3.2 The direct fabrication of monodisperse nanoparticles

In the previous section, the relationship between the growth rate and the size distribution of nanoparticles has been discussed. By understanding the growth process of nanoparticles, it is possible to control their size distribution. Laser Vaporization with Controlled Condensation (LVCC), as shown in **Figure 5** [41], is a method to control the average size of the nanoparticles by adjusting the conditions about nucleation and growth. By raising the temperature of the lower wall in ablation chamber with a heater and cooling down the upper wall with liquid nitrogen, the temperature gradient would cause the convection of ambient gas. When the vapor created by laser ablation drifts upward by the convection, the vapor becomes supersaturated near the upper cold wall, which results in nucleation

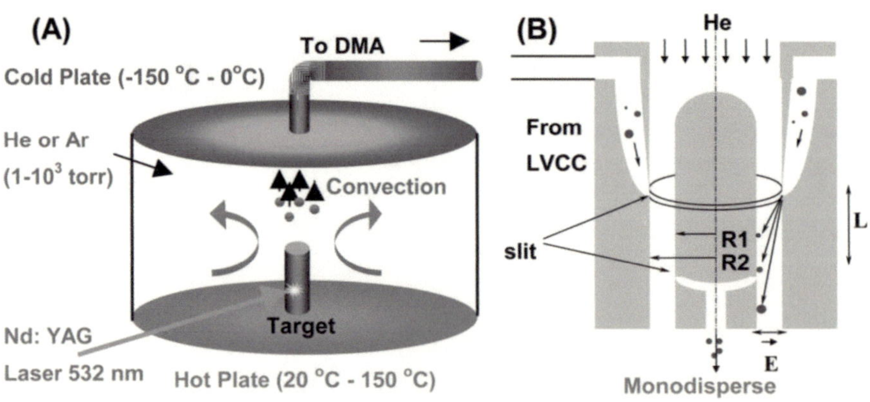

Figure 5.
(A) Experimental set-up for the synthesis of nanoparticles using the LVCC method. (B) Experimental setup for the LVCC method coupled with a DMA [41].

and condensation followed by nanoparticle forming. It is found that the higher the supersaturation is, that is, the larger the temperature gradient is, the smaller the average size of the nanoparticles becomes. Therefore, by adjusting the temperature gradient proficiently, it is possible to control the nanoparticle size. Furthermore, by combining LVCC and DMA, to control the average diameter and size distribution of nanoparticles would be possible.

In order to determine the size distribution by directly adjusting the growth process of the nanoparticles, it is necessary to control not only statically the degree of supersaturation but also dynamically its temporal variation of it. That is, it is necessary to rapidly increase the supersaturation level to complete the nucleation within a short period of time and to inhibit the subsequent nanoparticle growth in some ways [42].

As described in Section 2.3, a shock wave is generated in front of the plume by laser ablation. It is possible to use the shock wave to rapidly change the state quantity in the plume and increase the supersaturation at a fast rate. One example is the laser ablation process done in a closed space such as an ellipsoidal cell, in which the collision between the reflected shock wave and the plume front triggers the formation of nanoparticles instantly in a small area [43, 44]. The equipment fabricating monodispersed nanoparticles by using this phenomenon is called Spatiotemporal Confined Nanoparticle Source (SCCS), which is illustrated in **Figure 6** [45], as the fabrication process is restrained in the confined space and time.

3.3 Application of self-ordered nanoparticle films on substrate

The possibility of developing new functional materials with nanoparticle films has been pointed out for a long time [46], and many attempts have been made mainly to create light-emitting devices using the visible light emission from silicon

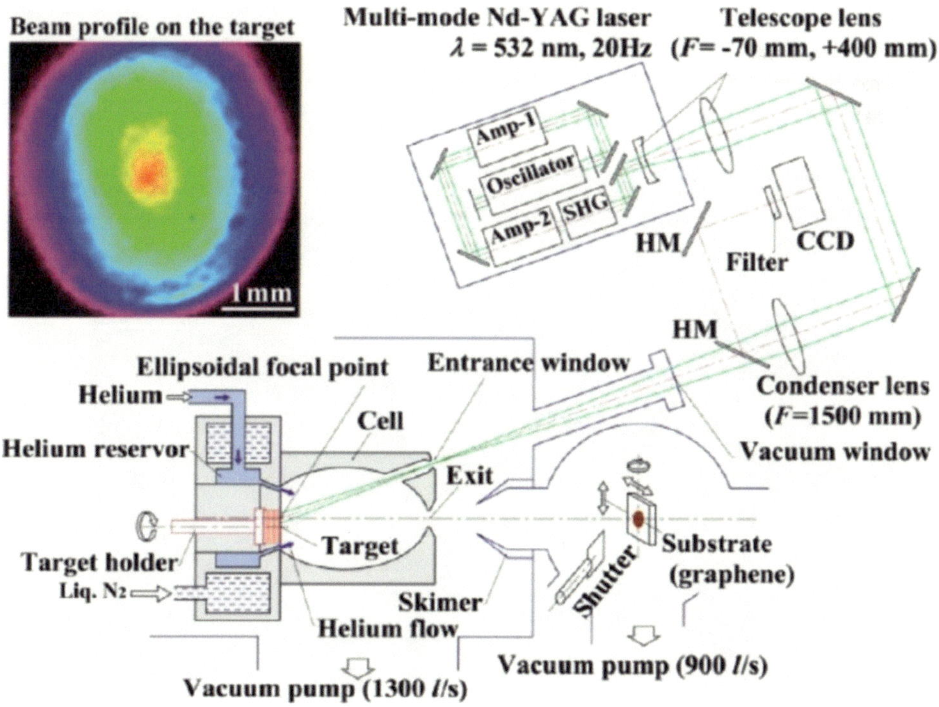

Figure 6.
Schematic view of new laser ablation-type silicon cluster beam system named SCCS [45].

nanoparticles [47–51]. Chen et al. [52] accumulated silicon nanoparticles on a substrate in an argon and oxygen atmosphere and researched on the effects of the gas pressure and annealing way on photoluminescence (PL). As a result, it was concluded that the emission band of 1.8–2.1 eV obtained from the experiments is based on the quantum effect since the blueshift of the emission varies depending on the nanoparticle size. Also, by accumulating silicon nanoparticles on a substrate in helium gas atmosphere, Kabashin et al. [53] found out that the microstructure on the nanoparticle film's surface varied depending on the helium gas pressure. When the helium gas pressure is higher than 1.5 Torr, it becomes a shape of porous film. Furthermore, it is concluded that the morphology of microstructure on the film surface determines the extent to which the natural oxidation of silicon nanoparticle affects PL luminescence.

In addition, the nanoparticle properties of iron oxide [54] and cobalt oxide [55] are studied for applying to a new functional device and material such as magnetic recording media, magnetic fluids and gas sensors. The properties of tungsten oxide nanoparticles [56] have been under research for being used as photocatalysts. However, while these applications of nanoparticles are still in the experimental stage, monodispersed nanoparticle beams are expected to improve the accuracy of these experiments.

On the other hand, from before, some studies focused attention on the function of laser irradiation to create an ordering of nanoparticles and investigated the influences of the intense and direction of electric field on the ordering appearance of nanoparticles. The phenomenon was called by Laser-Induced Periodic Surface Structures (LIPSS) [57, 58], which have led to a growing interest on how to build a new order of the nanoparticle array, as represented by the image of **Figure 7**. And Han et al. [59] confirmed by experiments and simulations that low-energy nanoparticle beams are effective in producing films of small nanoparticles with a wide range of sizes. Moreover, since the formation of ordered array is essential for making larger films, the technology of forming ordered nanoparticles by irradiating the substrate's surface with the beam of monodispersed nanoparticles which have a stable crystal structure has been expected. It is verified that the silicon nanoparticles fabricated by this technology spontaneously form a nanostructure ordering, and the formation

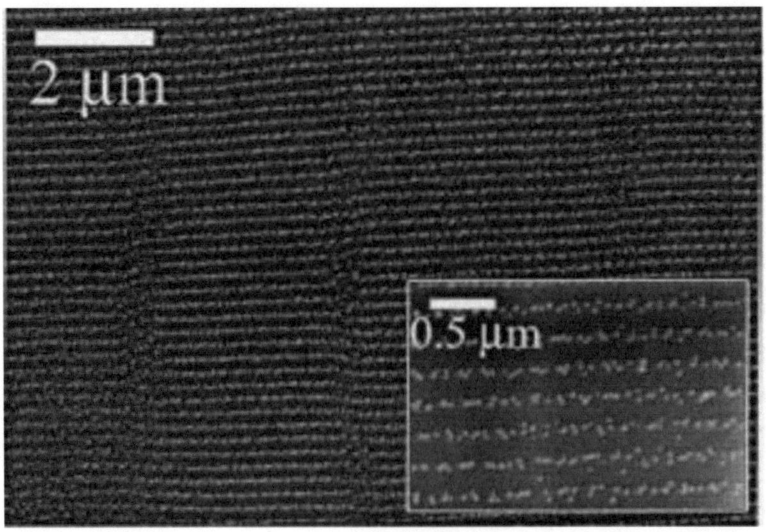

Figure 7.
SEM image of silicon nanoparticles produced by PLD and their long-range ordering controlled by subsequent laser irradiation [57].

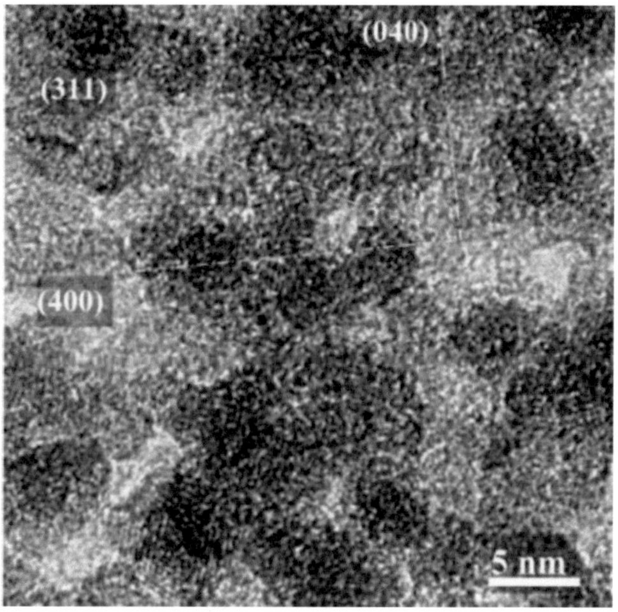

Figure 8.
HRTEM image of silicon cluster superlattice structures obtained with the use of the new laser ablation-type silicon cluster beam system [45].

mechanism has been studied [60]. The components of this structure are called silicon nanoblocks [61], which are expected to be applied on next-generation devices such as accumulator integration system with a solar cell [62], ultra-thin supercapacitor [63] and superlattice structures [45]. The silicon clusters possesing atomic crystalline structures generated by SCCS are depositing on the graphene substrate and forming shapes of silicon superlattice as shown in **Figure 8**.

In addition, it is found that interparticle spacing and pattern are the important parameters for characterizing the film consisting of nanoparticle array. We can extract the useful properties from the nanoparticle films, depending on the interparticle spacing range of electron tunneling, optical near field or spin exchange interaction. In the next generation of laser ablation, it seems to be one of critical challenges to clarify and control the ordering of nanoparticles based on the viewpoint of the interparticle spacing.

4. Summary

To apply the nanostructured particle films on new functional materials, it is important to study the formation process. Therefore, in this chapter, we stared from the mechanism of laser ablation process and then found that monodispersed nanoparticle beams are necessary for fabricating nanoparticle films and that, when used as a macroscopic material, the technology based on the self-ordering of nanoparticles is essential to make large films. It is expected to control the interparticle spacing and its pattern of nanoparticle structure in the films to obtain new and useful properties which are potentially underlying in the films as functional devices. In order to realize such devices, laser ablation remains a promising technology for the future, and worth studied for more possibilities.

Author details

Toshio Takiya[1,2]* and Naoaki Fukuda[1,3]

1 Hitachi Zosen Corporation, Osaka, Japan

2 Hitz Collaborative Research Institute, Osaka University, Osaka, Japan

3 Research Center for the 21st Century, Osaka Prefecture University, Osaka, Japan

*Address all correspondence to: takiya@hitachizosen.co.jp

References

[1] H. W. Kroto, J. R. Heath, S. C. O'Brien, R. F. Curl and R. E. Smalley, C60: Buckminsterfullerene, Nature, vol. 318, p.62-163 (1985)

[2] W. T. Nichols, G. Malyavanatham, D. E. Henneke, D. T. O'Brien, M. F. Becker and J. W. Keto, Bimodal Nanoparticle Size Distributions Produced by Laser Ablation of Microparticles in Aerosols, Journal of Nanoparticle Research, vol.4, p.423-432(2002)

[3] A. E. Mayer and A. A. Ebel, Shock-induced compaction of nanoparticle layers into nanostructured coating, J. Appl. Phys. 122, 165901 (2017)

[4] T. Takiya, N. Fukuda, N. Inoue, M. Han, M. Yaga and Y. Iwata, Dynamics of the Shock Wave Accompanied by Nanoparticle Formation in the PLA Processes, Adv. Studies Theor. Phys., vol.4-7, 305-316(2010)

[5] E. Ueno, N. Fukuda, H. Fukuoka, M. Yaga, I. Umezu, M. Han and T. Takiya, Nanoparticle formation by interaction between laser ablated plume and shock waves, 17th International Symposium on Small Particles and Inorganic Clusters, (2014)

[6] A. Higo, K. Katayama, H. Fukuoka, T. Yoshida, T. Aoki, M. Yaga and I. Umezu, Expansion of laser-induced plume after the passage of a counter shock wave through a background gas, Appl. Phys. A, vol.126, Article number: 304 (2020)

[7] A. Vertes, R. W. Dreyfus and D. E. Platt, Modeling the thermal-to-plasma transitions for Cu photoablation, IBM J. Res. Develop. 38-1, 4-10(1994)

[8] A. Peterlongo, A. Miotello and R. Kelly, Laser-pulse sputtering of aluminum: Vaporization, boiling, superheating, and gas-dynamic effects, Phys. Rev. E 50-6, 4716-4727(1994)

[9] F.A. Houle and W.D. Hinsberg, Stochastic simulation of heat flow with application to laser–solid interactions, Appl. Phys. A, vol.66, p.143-151(1998)

[10] M. Han, S. Kiyama, M. Muto, A. Fukuda, T. Sawada and Y. Iwata, Cluster formation dynamics in a locally-confined gas layer mixed with the plume ablated by pulsed laser irradiation, Nucl. Instr. Meth. Phys. Res., B153, 302-308(1999)

[11] S. I. Anisimov, Vaporization of Metal Absorbing Laser Radiation, J. Exp. Theor. Phys., 27-1, 182-183(1968)

[12] R. Kelly, On the dual role of the Knudsen layer and unsteady, adiabatic expansion in pulse sputtering phenomena, J. Chem. Phys. 92-8, 5047-5056(1990)

[13] R. Kelly, A. Miotello, A. Mele, A. G. Guidoni, J. W. Hastie, P. K. Schenck and H. Okabe, Gas-dynamic effects in the laser-pulse sputtering of AlN: is there evidence for phase explosion?, Appl. Surf. Sci., vol.133-4, p.251-269(1998)

[14] C. J. Knight, Theoretical Modeling of Rapid Surface Vaporization with Back Pressure, AIAA J, 17-5, 519-523(1979)

[15] L. V. Zhigilei and B. J. Garrison, Velocity distributions of molecules ejected in laser ablation, Appl. Phys. Lett. 71-4, 551-553(1997)

[16] A. V. Gusarov and I. Smurov, Target-vapour interaction and atomic collisions in pulsed laser ablation, J. Phys. D: Appl. Phys. 34, 1147-1156(2001)

[17] A. V. Bulgakov and N. M. Bulgakova, Gas-dynamic effects of the interaction between a pulsed laser-ablation plume and the ambient gas: analogy with an underexpanded jet, J. Phys. D: Appl. Phys. 31, 693-703(1998)

[18] S. S. Harilal, C. V. Bindhu, M. S. Tillack, F. Najmabadi and A. C. Gaeris,

Plume splitting and sharpening in laser-produced aluminium plasma, J. Phys. D: Appl. Phys. 35, 2935-2938(2002)

[19] I. Umezu, M. Inada, K. Kohno, T. Yamaguchi, T. Makino and A. Sugimura, Reaction between nitrogen gas and silicon species during pulsed laser ablation, J. Vac. Sci. Technol. A 21-5, 1680-1682(2003)

[20] Y. Nakata, G. Soumagne, T. Okada, M. Maeda, Pulsed-laser deposition of barium titanate films and plume dynamics, Appl. Surf. Sci., p.127-129, 650-654(1998)

[21] D. B. Geohegan, Fast intensified-CCD photography of YBa2Cu3O7−x laser ablation in vacuum and ambient oxygen, Appl. Phys. Lett. 60-22, 2732-2734(1992)

[22] D. B. Geohegan, A. A. Puretzky, G. Duscher and S. J. Pennycook, Time-resolved imaging of gas phase nanoparticle synthesis by laser ablation, Appl. Phys. Lett. 72-23, 2987-2989(1998)

[23] D. B. Geohegan, A. A. Puretzky, G. Duscher and S. J. Pennycook, Photoluminescence from gas-suspended SiOx nanoparticles synthesized by laser ablation, Appl. Phys. Lett. 73-4, 438-440(1998)

[24] M. Han, Y. Gong, J. Zhou, C. Yin, F. Song, N. Muto, T. Takiya and Y. Iwata, Plume dynamics during film and nanoparticles deposition by pulsed laser ablation, Phys. Lett. A302, 182-189(2002)

[25] O. F. Hagena and W. Obert, Cluster Formation in Expanding Supersonic Jets: Effect of Pressure, Temperature, Nozzle Size, and Test Gas, J. Chem. Phys. 56-5, 1793-1802(1972)

[26] for example, D. J. McGinty, Molecular dynamics studies of the properties of small clusters of argon atoms, J. Chem. Phys. 58-11, 4733-4742(1973)

[27] D. T. Gillespie, A stochastic analysis of the homogeneous nucleation of vapor condensation, J. Chem. Phys. 74-1, 661-678(1981)

[28] for example: N. Bykov and Y. Gorbachev, Mechanics - Seventh Polyakhov's Reading, 2015 International Conference, St.-Petersburg, Russia, February (2015)

[29] L. Landström, Formation of Nanoparticles by Laser-Activated Processes, Comprehensive Summaries of Uppsala Dissertations from the Faculty of Science and Technology 855, p.10, Acta Universitatis Upsalensis, Uppsala, (2003)

[30] J. Söderlund, L. B. Kiss, G. A. Niklasson, and C. G. Granqvist, Lognormal Size Distributions in Particle Growth Processes without Coagulation, Phys. Rev. Lett. 80-11, 2386-2388 (1998)

[31] L. B. Kiss, J. Söderlund, G. A. Niklasson and C. G. Granqvist, New approach to the origin of lognormal size distributions of nanoparticles, Nanotechnology 10, 25-28 (1999)

[32] N. Koshizaki, A. Narazaki and T. Sasaki, Size distribution and growth mechanism of Co3O4 nanoparticles fabricated by pulsed laser deposition, Scripta mater. 44, 1925-1928 (2001)

[33] T. Johnson, R. Caldow, A. Pöcher, A. Mirme and D. Kittelson, An Engine Exhaust Particle Sizer Spectrometer for Transient Emission Particle Measurements, 7th ETH- ETH Conference on Combustion Generated Particles, Zürich, Switzerland (2003)

[34] M. Kim, S. Osone, T. Kim, H. Higashi and T. Seto, Synthesis of Nanoparticles by Laser Ablation: A Review, KONA Powder and Particle Journal No. 34 (2017)

[35] P. M. Denby and D. A. Eastham, Efficient technique for producing

high-brightness, size-selected cluster beams, Appl. Phys. Lett. 79-15, 2477-2479 (2001)

[36] E. Hontañón and F. E. Kruis, A Differential Mobility Analyzer (DMA) for Size Selection of Nanoparticles at High Flow Rates, Aeros. Sci. and Technol., vol.43-1, 25-37(2009)

[37] T. Seto, Y. Kawakami, N. Suzuki, M. Hirasawa, S. Kano, N. Aya, S. Sasaki and H. Shimura, Evaluation of Morphology and Size Distribution of Silicon and Titanium Oxide Nanoparticles Generated by Laser Ablation, J. Nanoparticle Res., vol.3, p.185-191 (2001)

[38] R. P. Camata, H. A. Atwater, K. J. Vahala and R. C. Flagan, Size classification of silicon nanocrystals, Appl. Phys. Lett. 68-22, 3162-3164 (1996)

[39] N. Suzuki, T. Makino, Y. Yamada, T. Yoshida and T. Seto, Monodispersed, nonagglomerated silicon nanocrystallites, Appl. Phys. Lett. 78-14, 2043-2045 (2001)

[40] H. P. Wu, A. Okano, K. Takayanagi, Photoluminescence properties of size-selected Si nanocluster films prepared by laser ablation, Appl. Phys. A 71-6, 643-646(2000)

[41] G. Glaspell, V. Abdelsayed, K. M. Saoud, and M. S. El-Shall, Vapor-phase synthesis of metallic and intermetallic nanoparticles and nanowires: Magnetic and catalytic properties, Pure Appl. Chem., vol. 78-9, pp.1667-1689(2006)

[42] S. Yamamuro, K. Sumiyama and K. Suzuki, Monodispersed Cr cluster formation by plasma-gas-condensation, J. Appl. Phys., 85-1, 483-489 (1999)

[43] Y. Iwata, M. Kishida, M. Muto, S. Yu, T. Sawada, A. Fukuda, T. Takiya, A. Komura and K. Nakajima, Chem. Phys. Lett. 358, 36-42(2002)

[44] Y. Iwata, M. Muto, T. Sawada, M. Han, A. Fukuda, S. Okayama, H. Matsuhata, H. Yamauchi, M. Kishida, T. Takiya, A. Komura and K. Nakajima, Well-defined Cluster Beam Deposition (CBD) Developed for Vacuum Synthesis of Nanostructures, Proc. 5th ISTC Scientific Advisory Committee Seminar, p.1-6, St Petersburg, Russia (2002)

[45] Y. Iwata, K. Tomita, T. Uchida and H. Matsuhata, Crystallographic Coalescence of Crystalline Silicon Clusters into Superlattice Structures, Cryst. Growth Des. 15-5, 2119-2128(2015), DOI: 10.1021/cg5016753

[46] G. D. Stein, DESIGN AND USE OF METAL CLUSTER BEAM SOURCES: IMPLICATIONS FOR THIN FILM DEVICES, Proc. Int. Ion Eng. Cong., ISIAT'83 & IPAT'83, Kyoto, 1165-1176B (1983)

[47] S. Li, S. J. Silvers and M. S. El-Shall, Surface Oxidation and Luminescence Properties of Weblike Agglomeration of Silicon Nanocrystals Produced by a Laser Vaporization−Controlled Condensation Technique, J. Phys. Chem. B 101, 1794-1802 (1997)

[48] L. Patrone, D. Nelson, V.I. Safarov, S. Giorgio, M. Sentis & W. Marine, Synthesis and properties of Si and Ge nanoclusters produced by pulsed laser ablation, Appl. Phys. A69[Suppl.], S217-S221 (1999)

[49] F. Huisken, B. Kohn and V. Paillard, Structured films of light-emitting silicon nanoparticles produced by cluster beam deposition, Appl. Phys. Lett. 74-25, 3776-3778 (1999)

[50] L. Patrone, D. Nelson, V. I. Safarov, M. Sentis and W. Marine, Photoluminescence of silicon nanoclusters with reduced size dispersion produced by laser ablation, J. Appl. Phys. 87-8, 3829-3837 (2000), DOI: 10.1063/1.372421

[51] A. V. Kabashin and M. Meunier, Photoluminescence characterization of Si-based nanostructured films produced by pulsed laser ablation, J. Vac. Sci. Technol. B 19-6, 2217-2222 (2001), DOI: 10.1116/1.1420494

[52] X. Y. Chen, Y. F. Lu, Y. H. Wu, B. J. Cho, M. H. Liu, D. Y. Dai and W. D. Song, Mechanisms of photoluminescence from silicon nanocrystals formed by pulsed-laser deposition in argon and oxygen ambient, J. Appl. Phys. 93-10, 6311-6319(2003), DOI: 10.1063/1.1569033

[53] A. V. Kabashin, J.-P. Sylvestre, S. Patskovsky and M. Meunier, Correlation between photoluminescence properties and morphology of laser-ablated Si/SiOx nanostructured films, J. Appl. Phys. 91-5, 3248-3254(2002), DOI:10.1063/1.1446217

[54] L. Zbroniec, T. Sasaki, N. Koshizaki, Ambient gas effects on iron oxide particle aggregated films prepared by laser ablation, Scripta mater. 44, 8-9, 1869-1872(2001), DOI: 10.1016/S1359-6462(01)00735-7

[55] Q. Li, T. Sasaki and N. Koshizaki, Pressure dependence of the morphology and size of cobalt (II,III) oxide nanoparticles prepared by pulsed-laser ablation, Appl. Phys. A 69, 115-118(1999), DOI: 10.1007/s003390050982

[56] S. Li, M. S. El-Shall, Synthesis of nanoparticles by reactive laser vaporization: silicon nanocrystals in polymers and properties of gallium and tungsten oxides, Appl. Surf. Sci. 127-129, 330-338(1998), DOI: 10.1016/S0169-4332(97)00651-X

[57] J. D. Fowlkes, A. J. Pedraza, D. A. Blom and H. M. Meyer III, Surface microstructuring and long-range ordering of silicon nanoparticles, Appl. Phys. Lett. 80-20, 3799-3801(2002), DOI: 10.1063/1.1480106

[58] A. J. Pedraza, J. D. Fowlkes, D. A. Blom, H. M. Meyer, Laser-induced nanoparticle ordering, J. Mater. Res. 17-11, 2815-2822(2002), DOI: 10.1557/JMR.2002.0409

[59] Han Min, Wang Zhaoye, Chen Pingping, Yu Shengwen, Wang Guanghou, Mechanism of neutral cluster beam deposition, Nucl. Instr. Meth. Phys. Res. B135, 1-4, 564-569(1998), DOI: 10.1016/S0168-583X(97)00635-6

[60] M. Muto, M. Oki, Y. Iwata, H. Yamauchi, H. Matsuhata, S. Okayama, Y. Ikuhara, T. Iwamoto and T. Sawada, Silicon nanoparticle lattice system(CLS) formed on an amorphous carbon surface by supersonic nanoparticle beam irradiation, Proc. Int. Symp. On Atomic Nanoparticle Collisions, Saint-Peteraburg(2003)

[61] Y. Iwata, Silicon Nanoblocks Pave The Way for A New Conceptual Nanoarchitecture, AIST Today Vol. 3, No. 7, 4-6 (2003)

[62] Q. Rehman, A. D. Khan, M. Noman, H. Ali, A. Raufd and M. S. Ahmad, Super absorption of solar energy using a plasmonic nanoparticle based CdTe solar cell, RSC Advances, Issue 59, (2019), DOI: 10.1039/C9RA07782K

[63] M. Hassan, M. A. Gondal, E. Cevikc, T. F. Qahtan, A. Bozkurt, M. A. Dastageer, High performance pliable supercapacitor fabricated using activated carbon nanospheres intercalated into boron nitride nanoplates by pulsed laser ablation technique, Arabian J. Chem., vol.13-8, p.6696-6707(2020), DOI: 10.1016/j.arabjc.2020.06.024

Section 3

Plasma Generated by Laser Ablation

Chapter 7

Dynamics of Transient Plasmas Generated by ns Laser Ablation of Memory Shape Alloys

Stefan Andrei Irimiciuc, Norina Forna, Andrei Agop, Maricel Agop, Stefan Toma and Doriana Forna Agop

Abstract

Understanding the underline fundamental mechanism behind experimental and industrial technologies embodies one of the foundations of the advances and tailoring new materials. With the pulsed laser deposition being one of the key techniques for obtaining complex biocompatible materials with controllable stoichiometry, there is need for experimental and theoretical advancements towards understanding the dynamics of multi component plasmas. Here we investigate the laser ablation process on Cu-Mn-Al and Fe-Mn-Si by means of space-and time-resolved optical emission spectroscopy and fast camera imaging. In a fractal paradigm the space–time homographic transformations were correlated with the global dynamics of the ablation plasmas.

Keywords: shape memory alloy, laser ablation, transient plasma, optical emission spectroscopy, fractal model

1. Introduction

The dynamics of the ejected particles as a results of high power laser and solid matter is not a trivial problem, as it was showcased in several papers [1, 2]. The problem of complex materials, as it is the case of metallic alloys, it consists in differences in the physical properties of the composing elements. Phenomena like heterogenous melting and vaporization [3] are commonly reported for ns laser ablation, with dire consequences for applications like pulse laser deposition. Target material heterogeneity should be reflected in the dynamics of the ejected particles, which is often difficult to observe in industrial applications like laser welding, cutting, surface cleaning, but is otherwise excellent showcased in applications like LIBS or plasma spectroscopy. The amalgam of plasma entities found in a transient plasma generated by laser ablation contains ions, atoms, molecules, electron and photons. The most often used technique extensively reported by the other groups [4] or even by our group are non-invasive ones that can differentiate between the contribution of each individual component of the plasma in particular conditions even reflect the complex local and global phenomena reported in recent years. These techniques are mainly concerning the optical emission spectroscopy. Understanding laser based technologies and the interaction between high energy laser beam and metallic alloys are now relevant for a wide range of applications with fast feedback and accurate predictions on the behavior of physical processes. The dual approach of experimental investigations and theoretical modeling has proven to be a successful method for understanding the dynamics of multi-element fluids [5, 6] or as it was showcased recently by our group for complex laser produced plasmas (LPP) [7]. The study presented in this chapter expands our previous attempts for stoichiometric transfer and plasma chemistry in the case of laser ablation of complex alloys. We discuss here the ablation of metallic particles as a result of short laser ablation interaction with ternary alloys from both an experimental and theoretical point of view. To comprehend the ablated particle dynamics we implemented optical emission spectroscopy in conjecture with ICCD fast camera imaging to record global and local information about their spatial distribution within the ablated cloud and their individual kinetic and thermal energy. From a theoretical perspective we built on our model from [7] and focused on exploring under, a fractal paradigm of motion, the effect of the plasma thermal energy (temperature) and ion physical properties (mass) on the spatial distribution of complex alloy plasmas. Usual models used to simulate the dynamics of complex systems are based on an assumption of the physical variable differentiability (e.g. density, momentum, energy, etc. [8–12] and the processes which they define. The practicality of such methods can be accepted sequentially, on space–time domains for which the differentiability still respected. However, the differential approach often fails when confronted by the reality of complex physical system (i.e. plasma plume expansion in PLD). To better represent most of the interactions at both local and global scales, it is required to introduce explicitly the scale resolution dependence. This breathes a new physical system where the variable dynamic that previously were dependent only on space and time, will now contain explicitly the dependence on the scale resolution. This can be even more abstracted and instead of using non-differential function, admittedly rather difficult to implement, just utilize different approximations of these multifractal mathematical functions derived by means of averaging at various scale resolution. A paramount consequence of this approximation is that any dynamic variable will behave as a limit of specific function families, which are non-differentiable for a null scale resolution (multifractal functions).

2. Laser induced plasmas on memory shape alloys

When investigating the ejected cloud of particle, the *ideal* investigation technique should be non-invasive and offer global and local information about the plasma components. Such a technique can be considered the combination of ICCD fast camera imaging and space and time resolved optical emission spectroscopy, which is used consistently by our group [7, 13, 14] and it was also validated by a significant number of papers [15, 16]. Our approach was a global - local one, meaning that the initial step was to collect the global emission (in the 300–700 nm regime) of the LPP at consecutive time-delays with respect to the laser beam. The result for a Fe-Mn-Si laser produced plasma is shown in **Figure 1**, where we present selected images during the LPP expansion in a 2 μs time lapse.

We observe that the plasma has a quasi-spherical shape and increases its volume as plasma evolves. The expansion velocity was estimated using the technique presented other previous papers [1], where it is discussed the effect of multi-element composition of the ablation process. When performing cross-section on the recorded images in axial and transversal directions, we notice different behaviors across the two directions. We also performed cross section across the main expansion axis (axial cross-section) reveals a splitting of the plasma cloud in multiple

Figure 1.
Global evolution of the Fe-Mn-Si LPP.

distinct structures (two or three). Some studies report on a specific terminology for these structures, the *first one* also named the *fast one* is created by electrostatic mechanism (Cloulomb Explosion), the *second one* or *slower* structure is generated by thermal mechanisms (Explosive boiling), while the third consists of mainly clusters or nanoparticles. Their presence has beforehand been reported and extensively discussed in conjecture with the multiple ejection mechanism and their correlation with the fractality of the LPP [7, 14, 15, 17] by our group. However, our focus will not be on this third structure as the main optical signatures, seen through our experimental methods, are given by the dynamics of simpler plasma entities like atoms or ions. The velocities of the main structures were determined as follows: for the case of Cu-Mn-Al plasma – 15 km/s for the first structure and 7.4 km/s and for the second structure for the case of Fe-Mn-Si plasma - 20 km/s for the first one and 11 km/s for the second structure. The values are in good agreement with the other reports from literature [1, 18, 19]. The obtained values strongly are related to the differences of the melting points for each material and the overall mass of the cloud, with significant variance in the properties of the component directly affecting the ablation process and the subsequent evolution.

We notice a significant difference in the overall emission and shape of the LPP generated on the two alloys. The global emission is noticeably larger for the Fe-Mn-Si plasma and with less inner structuring, while for the Cu-Mn-Al the global emission is reduced and presents more pronounced structuring. These differences are induced by the energetic distribution uniformity on the excitation process as opposed to other types of interactions (i.e. ionization). Fe-Mn-Si plasma has an uniform aspect which is attributed similarities in the melting points of the composing elements, which leads to a uniform and homogeneous ablation. For the Cu-Mn-Al plasma there are significant differences between the physical properties of Al and Mn or Cu, could lead to a more heterogenous ablation process. These statements will further be verified with the space and time resolved OES. We would like to also note that, the fractality of the laser produced plasmas will also be affected by the inner energy of the plasma and its distribution on the composing entities [7, 13]. We anticipate here another type of analysis (fractal analysis) which we will further use in this study, that could offer valuable information about the laser produced plasmas.

In **Figure 2** we plotted the spatial distribution of atoms (Fe and Mn) from the Fe-Mn-Si plasma highlighting the discrepancies amongst the two elements. We would like to note that Si was not considered as the emission line intensity for its species insignificant (lower) than those of the other elements. The Fe atoms have a dual peak distribution, while the Mn one presents only a single peak distribution. This reads as Fe atoms can be excited throughout the whole plasma volume, especially at longer distance where the electron density is significantly lower. This assessment can also explain the elevated T_{ex} reported earlier and it is in line with the multi-structure scenario seen by fast camera photography (ICCD fast camera

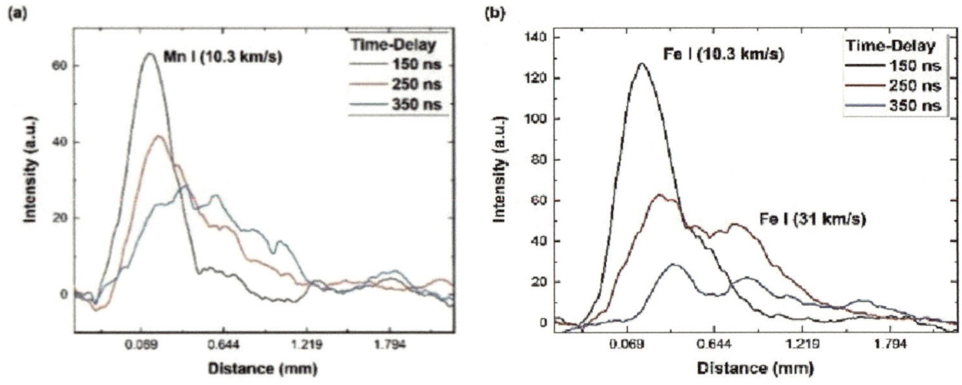

Figure 2.
Axial distribution of Fe (a) and Mn (b) atomic emission at various time-delays.

imaging). By representing the intensity maximum for individual emission line as a function of time [20], we determined the expansion velocities for the individual elements, with 31 km/s for the first peak of Fe I and 10 km/s for the second one, while for the Mn I a velocity of 18 km/s. The expansion velocity estimated for the first peak of Fe I and the Mn atoms are similar with the values of the second plasma structure, while the velocities of the second group of Fe atoms are in line with the value determined for the first plasma structure. This concludes the fact that the two-plasma structure have uniformly distributed atoms and ions amongst them with the fast structure having a slight depletion of Mn.

We can take a broader view of the discussions made in the previous paragaphs for both investigated plasmas as the laser fluence and background pressure and are (expansion conditions) identical. The results are seen in **Figure 3**-right-hand side, where we can observe for a time-delay of 150 ns the spatial distribution of Fe and Mn in the Fe-Mn-Si plasma and Cu an Al in the Cu-Mn-Al plasma, respectively. We notice that for lighter elements we obtain a narrow spatial distribution, while the *heavier* ones (Cu and Fe) have a wider distribution. These differences can be seen as a separation of the composing elements based on their physical properties. The separation was previously discussed by our group in [7, 13] where the fractality of the components played a significant role, based on that the spatial distributions of different elements are reflecting the elevated degree of fractality. Lighter elements will have a higher collision rate and thus a higher fractality degree, whereas the

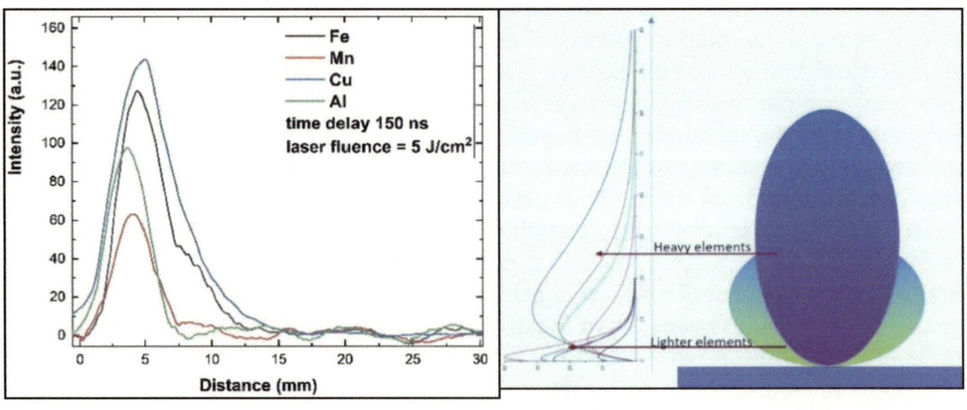

Figure 3.
Axial distribution of the main elements in the alloys as seen through OES measurements at a time-delay of 150 ns (left) and a schematic representation of the particle distribution with the plasma volume (right).

heavier ones are described by a lower fractality degree (lower collision rate). This difference in the fractality of the plasma entities will give us different spatial distributions for each element.

However, given our set-up optical configuration, lighter elements strongly scattered during expansion will appear to have a narrower distribution at relative short distances, while heavier particles will have a broader distribution most likely covering the whole plasma plume. Translating these results into the expansion of a three-dimensional plasma, low-mass elements are scattered towards the edge of the plasma plume while the high-mass ones are the building blocks the plasma core. For industrial applications like PLD, the result is of paramount importance interest as the particular volumes of the plasma plumes lack stoichiometry or uniformity. These properties could induce a non-congruent transfer of multielement material and affect the physical properties of the subsequent thin film. Furthermore, the diagnostic system used here allowed to capture the complex nature of the plasma and present some meaning behind it. We will further attempt to unravel more information about the relation between the fractality of specific elements and their spatial distribution within the plasma volume in the following section.

3. Theoretical modeling

The fractal analysis approach for understanding the dynamics of complex physical systems was shown over the years to provide with some of the most promising results towards understanding multiparticle flow in fluids [21, 22] or plasmas [7, 13–14, 17].

For a laser ablation plasma, the nonlinearity and the chaoticity have a dual applicability being both structural and functional, with the interactions between the so-called plasma entities (structural components like electrons, ions, atoms, photons) determine reciprocal conditioning micro–macro, local–global, individual-group, etc. In such a case, the universality of the laws describing the laser ablation plasma dynamics becomes obvious and it must be reflected by the mathematical procedures which are utilized. Basically, it makes use more and more often of the "holographic implementation" in the description of plasma dynamics. Usually, the theoretical models used to describe the ablation plasma dynamics are based on a differentiable variable assumption. Most of the notable results of the differentiable models must be understood sequentially, where the integrability and differentiability still apply. The differentiable mathematical procedures are limiting our understanding of more complex physical phenomena, such as the expansion of a laser produced plasma which implies various nonlinear behaviors, chaotic movement and self-structuring. In order to accurately describe the LPP dynamics and still remain tributary to differentiable and integral mathematics we must explicitly introduce the scale resolution. The scale resolution will be integrated in the expression of the physical variable, which describe the LPP, and implicitly in the fundamental equations, which govern these dynamics. This means that any physical variable becomes dependent on both spatial and temporal coordinates and the scale resolution. In other words, instead of using physical variables described by a nondifferentiable mathematical function, we will use different approximations of this mathematical function obtained through its averaging at various scale resolutions. As a consequence, the physical variables used to describe the LLP dynamics will act as a limit of functions family, which are non-differentiable for a null scale resolution and differentiable for non-null scale resolution.

This approach for describing LPP dynamics infers the building of novel geometric structures [23, 24] and probably new physical theories, in which the movement laws invariant to spatio-temporal transformation, can be considered integrated on scale laws, invariant to scale resolution conversions. These geometric structures can be generated by the multifractal theory of movement in the form of Scale Relativity Theory (SRT) with a fractal dimension $D_F = 2$ [25] or in the form of SRT in an arbitrary fractal dimension [21, 22]. In both cases the "*holographic implementation*" of specific dynamics of LPP suggests a substitution of dynamics with limitations in an Euclidian space with dynamics without any restriction in a free-multifractal space. Thus, we will use only of the expansion of the plasma particles on continuous and non-differentiable curves in a multifractal space [25].

In the following we will analyze some specific dynamics of a transient plasma generated by laser ablation, therefore postulating that the plasma particles are moving on multi-fractal curves. The mathematic procedure implies the usage of the following set of multifractal hydrodynamics equations. In such a context let us consider the density current:

$$\Im(x,t,dt) = \rho(x,t,dt)V(x,t,dt)\sum = \frac{\sum}{\pi^{1/2}} \frac{V_0\alpha^2 + \frac{4\lambda^2(dt)^{\frac{4}{F(\sigma)}-2}}{\alpha^2}xt}{\left[\alpha^2 + \frac{4\lambda^2(dt)^{\frac{4}{F(\sigma)}-2}}{\alpha^2}t^2\right]^{3/2}} \exp\left[-\frac{(x-V_0t)^2}{\alpha^2 + \frac{4\lambda^2(dt)^{\frac{4}{F(\sigma)}-2}}{\alpha^2}t^2}\right], \tag{1}$$

where Σ is a surface which $\Im$ crosses, the other parameters have the meaning given in [21, 22].

In the aforementioned conditions, $\Im$ is invariant with respect to the coordinates transformation group and to the scale resolutions transformation group. Since these two groups are isomorphs, between them we can unravel various isometries like: compactizations of the spatial and temporal coordinates, compactization of the scale resolutions, compactizations of the spatio-temporal coordinates and scale resolutions, etc. Following this we can perform a compactization between the temporal coordinate and the scale resolution, which is given by the relation:

$$\varepsilon = \frac{E}{m_0} = 2\lambda(dt)^{\frac{2}{F(\sigma)}-1}\upsilon, \quad \upsilon = \frac{1}{t}, \tag{2}$$

where ε corresponds to the specific energy of the ablation plasma entities. Once admitted such an isometry by means of substitutions:

$$I = \frac{\Im\pi^{1/2}\alpha}{V_0\sum}, \quad \xi = \frac{x}{\alpha}, \quad u = \frac{\varepsilon}{\varepsilon_0}, \quad \varepsilon_0 = \frac{2\lambda V_0(dt)^{\frac{2}{F(\sigma)}-1}}{\alpha}, \quad \mu = \frac{2\lambda(dt)^{\frac{2}{F(\sigma)}-1}}{\alpha V_0}, \tag{3}$$

(1) takes the more simplified non-dimensional form:

$$I = \frac{1+\mu^2\frac{\xi}{u}}{\left(1+\mu^2\frac{\xi}{u}\right)^{3/2}} exp\left[-\frac{\left(\xi-\frac{1}{u}\right)^2}{1+\left(\frac{\mu}{u}\right)^2}\right]. \tag{4}$$

In (5.3) and (5.4) I is assimilated to the normalized state intensity, ξ to the normalized spatial coordinate, μ to the normalized multifractalization degree, u to the normalized specific energy of the ablation plasma structures. The energy ε and the reference energy ε_0 can be written as:

$$\varepsilon \approx \frac{T}{M}, \varepsilon_0 \approx \frac{T_0}{M_0}, \tag{5}$$

where T and T_0 being the specific temperatures and M and M_0 the specific mass, we can further note:

$$\tau = \frac{T}{T_0}, \quad \theta = \frac{M}{M_0}. \tag{6}$$

so that (5.1) becomes:

$$I = \frac{1 + \mu^2 \xi \frac{\theta}{\tau}}{\left(1 + \left(\mu \frac{\theta}{\tau}\right)^2\right)^{3/2}} \exp\left[-\frac{\left(\xi - \frac{\theta}{\tau}\right)^2}{1 + \left(\mu \frac{\theta}{\tau}\right)^2}\right]. \tag{7}$$

The fundamental transient plasmas dynamics induced by laser ablation can be corelated with a multifractal medium for which its fractality degree is echoed by the elementary processes (collision, excitations, ionization or recombination, etc. -for other details see [7, 17]). In such a context (1) defines both the normalized state intensity and it is also measure of the optical emission of each plasma structure, case for which its spatial distribution of mass type is quantified through our mathematical model and corelated with our data.

The results of our simulations are presented in **Figure 4(a,b)**. One can see that plasma entities with a fractality degree $\mu < 1$ are defined by a narrow distribution centered around small values of ξ, while for a fractality degree $\mu > 1$ the distribution is wider and is centered around values of one order of magnitude higher that the low fractality ones. Therefore, we can formulate an particular image of the plasma plume dynamics in a multifractal mathematical formalism as follows: a core of entities with low fractality and a relative low plasma temperature as well a *shell* of high energetic particles described by a higher fractality degree.

In order to perform some comparison between our results and find if they can be correlated with the classical view of the LPP we have effectuated supplementary simulations on the plasma emission distribution over the particles mass for a plasma with an overall μ factor of 5 at an arbitrary distance (ξ = 5.5). The plasma entities with a lower mass are described by a higher relative emission for a particular temperature, and with the increase of the plasma temperature the emission of high-mass elements increases as well. The obtained data is in accordance with some of our previous results from [7, 13, 17], where we correlated the plasma temperature with the plasma fractal energy. These results can have real implications for some technological applications: for low plasma excitation temperature the distribution is

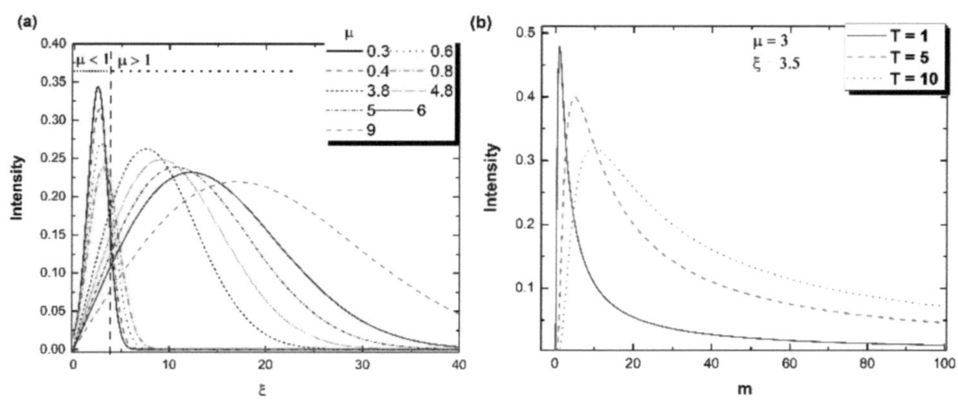

Figure 4.
Spatial dependence of the simulated optical emission of plasma entities with various fractal degrees (a) and mass distribution of the optical emission for various plasma temperature (b).

strongly heterogenous and it aides particles with an elevated fractalization degree leading to a non-congruent transfer in case of PLD and the lighter elements are predominantly in the outer regions of the plume, while the heavier ones are mainly part of the core.

4. A multifractal theoretical approach for understanding the separation of particle flow during pulsed laser deposition of multicomponent alloys

The details of the model have been previously reported in [14]. Let us consider that the evolution of the plasma components (plasma entities) is defined by continuous but non-differential curves, in specific rage of values. This premits us to corelate the properties of plasma plume in a multifractal matrix and thus reducing the dynamics of the individual entities by integrating them with their respective multifractal trajectories (geodesics). Therefore, at extreme times scales with respect to the inverse of the maxim Lyapunov exponent [23], the classical trajectories (deterministic) are replaced by fractal geodesics (families of potential trajectories and the notion of defined spatial coordinates is replaced by that of probability densities.

In such a context, in agreement with the results from [21, 22] at a differentiable resolution scale the ablation plasma dynamics are driven by the specific fractal force:

$$F_F^i = \left[u_F^l + \frac{1}{4}(dt)^{(2/D_F)-1}D^{kl}\partial_k\right]\partial_l u_F^i \tag{8}$$

The introduction of this multifractal force in explicit manner is essential at and is responsible for the structuring of ablation plasma on each component, though a special velocity field. The functionality of our differential system of equations is given by:

$$F_F^i = \left[u_F^l + \frac{1}{4}(dt)^{(2/D_F)-1}D^{kl}\partial_k\right]\partial_l u_F^i = 0 \tag{9}$$

$$\partial_l u_F^l = 0 \tag{10}$$

(9) represents that at a differential scale resolution the multifractal force becomes null, while (10) represents the state density conservation law at non-differentiable scale resolution.

Generally speaking it is rather difficult to obtain an analytic solution for the system of equations considered here, taking into account its multifractal nature (through $u_F^l\partial_l u_F^i$ the multifractal convection and $D^{kl}\partial_l\partial_k u_F^i$ the multifractal type dissipation); also the fractalization type, introduced through multifractal type tensor D^{kl}, is left unknown purposefully in this particular representation of the model.

The continuous development of our multifractal model and its implementation for the simulation of *real* plasma like phenomena implies the definition of a three-dimensional plasma like-fluid flow of a with a revolution symmetry around the *z* axis, and investigate its dynamics through a 2-dimensional projection of the plasma in the (x,*y*) plane.

Considering the symmetry plane (*x*,*y*), the (9) and (10) system becomes:

$$u_{F_x}\frac{\partial u_{F_x}}{\partial x} + u_{Fx}\frac{\partial u_{F_x}}{\partial y} = \frac{1}{4}(dt)^{(2/D_F)-1}\,D^{yy}\frac{\partial^2 u_{F_x}}{\partial y^2} \tag{11}$$

$$\frac{\partial u_{F_x}}{\partial x} + \frac{\partial u_{F_y}}{\partial y} = 0 \tag{12}$$

Let us solve the equation system (11) and (12) by imposing the following conditions

$$\lim_{y \to 0} u_{F_y}(x, y) = 0, \lim_{y \to 0} \frac{\partial u_{F_x}}{\partial y} = 0, \lim_{y \to \infty} u_{F_x}(x, y) = 0 \tag{13}$$

$$\Theta = \rho \int_{\cdot\cdot}^{+\infty} u_x{}^2 dy = \text{const.}$$

with:

$$D^{yy} = a \exp(i\theta) \tag{14}$$

Let us highlight that existence of a complex phase can be the pathway to a hidden temporal evolution of the system. The variation of a complex phase defines a time-dependence in an implicit manner. This means that for multifractal system can describe both spatial and temporal evolutions. Thus, the choice for D^{yy} gives the possibility of a both spatial and temporal investigations on the LPP plasma dynamics.

The solution of (11) and (12), in their general form in normalized quantities can be written as:

$$X = \frac{x}{x_0}, Y = \frac{y}{y_0}, U = u_{F_x}\frac{4y_0^2}{x_o a}, V = u_{F_y}\frac{4y_0^2}{x_o a}, \frac{\left(\frac{\Phi_0}{6\rho}\right)^{1/3}}{\left(a/4\right)^{2/3}} = \frac{x_0^{2/3}}{y_0}, \mu = (\mathrm{dt})^{(D_F/2)-1} \tag{15}$$

is given according to the method from [21, 22]:

$$U(X, Y) = \frac{3/2}{[\mu X]^{1/3} \exp\left(\frac{i\theta}{3}\right)} \cdot \mathrm{sech}^2\left\{\frac{1/2 Y}{[\mu X]^{2/3} \exp\left(\frac{2i\theta}{3}\right)}\right\} \tag{16}$$

$$V(X, Y) = \frac{(9/2)^{2/3}}{[\mu X]^{1/3} \exp\left(\frac{i\theta}{3}\right)}\left\{\left[\frac{Y}{\mu X^{2/3} \exp\left(\frac{2i\theta}{3}\right)}\right] \cdot \mathrm{sec}\,h^2\left[\frac{1/2 Y}{[\mu X]^{2/3} \exp\left(\frac{2i\theta}{3}\right)}\right] - \tanh\left[\frac{1/2 Y}{[\mu X]^{2/3} \exp\left(\frac{2i\theta}{3}\right)}\right]\right\} \tag{17}$$

To verify the validity of such unusual approach we obtained 3D (**Figure 5**) representations of the transient plasma flow developed based on the solution given by our multifractal system of equations. The transient plasma is *generated* in the framework of our multifractal model as a mixture of various particles with different physical properties (electron, ions, atoms, nanopatricles). This implies that the some multifractal parameters such as the complex phase, fractal dimension or specific length (x_0, y_0) will include within their values the properties of each individual component. In **Figure 5** we represented the angular separation of the plasma low for different values of the complex phase leading to the appearance of preferential *expansion directions for various elements of the plasma* for $\Theta > 1.5$.

In **Figure 6** we have represented the 2D distribution portraying various plasma flow scenarios with respect to the structure of the laser ablation plasma, starting form a pure, single ionized plasma (only atoms, ions and electrons) towards a multi-component flow (including nanoparticles, molecules or clusters). There is a separation into multiple structures in the two expansion directions (across *X* and *Y*).

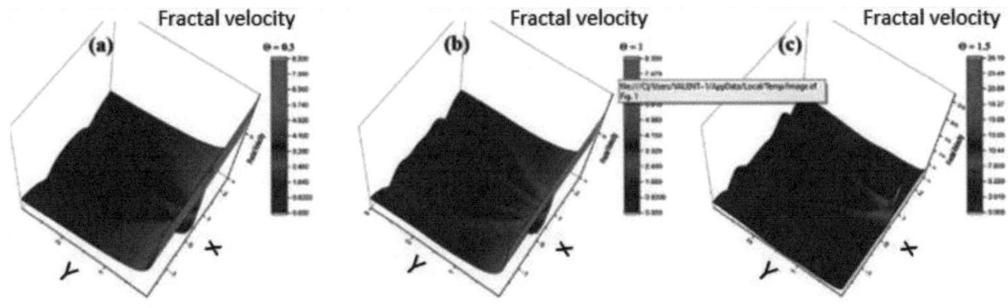

Figure 5.
3D representation of the total fractal velocity field of a multi-fractal plasma flow for various complex phases (0.5 – (a) 1 – (b) 1.5 – (c)).

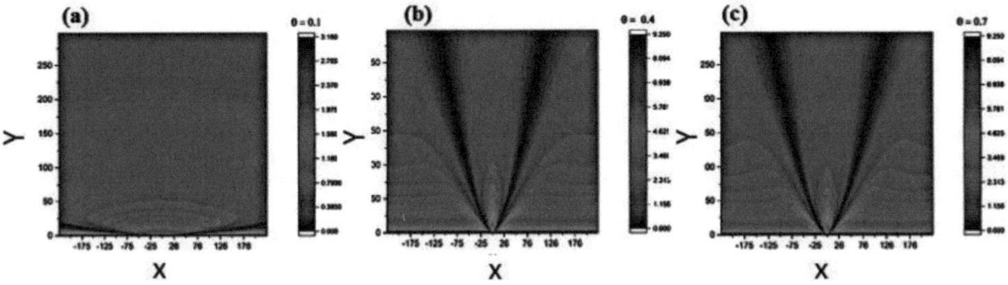

Figure 6.
Total fractal velocity field evolution on the two main directions (X,Y) for a multi fractal system with ξ values of 0.1 (a), 0.4 (b), 0.7 (c).

For small values of the fractalization degree, which from here on further will be considered as *control* parameter, we define a plasma flow containing only one type of particles thus on plasma component. In **Figure 6(a)**, there can be seen only one plasma structure along the main expansion axis. The increase of this fractalization degree, control parameter, subsequently leads to changes in the homogeneity of the structural units of the plasma (i.e. our model of plasma becomes more heterogeneous in term of plasma particle mass and energy). It is also noticeable the formation of two symmetrically positioned secondary structures (lateral with respect to the main expansion axis). We conclude that those volume of plasma contain mainly components defined by a small physical volume and low kinetic energy. A subsequent increase in the fractalization degree and thus of the heterogeneity of the plasma leads to the formation of lateral-symmetrically situated plasma *structures* each defining different families of particles with physical properties.

An important conclusion extracted from our simulation is that the plasma structuring process is gradual one. For values of $\xi = 0.3 \sim 1$, we can obtain three main lateral structures which are also followed by a continuous internal structuring visible for fractalization degrees $\xi > 1$. Within the framework of our multifractal model this is reversible transition as the distribution often returns to the three-structure system. Another approach of understating this novel phenomenon is to assimilated then with *breathing modes of the fractal system* (oscillatory behavior). The evolution of our plasma model in a multifractal framework attempts a complete transition towards a completely separate flow but the interactions of the multifractal forces between the individual plasma structures are then in charge for the unification of the plasma structures.

The multi-structuring of the laser produced plasma was highlighted by executing cross sections in X direction (see **Figure** 7). In X direction the separation is more obvious first moments of expansion. Each of the new plasma structure is defined by

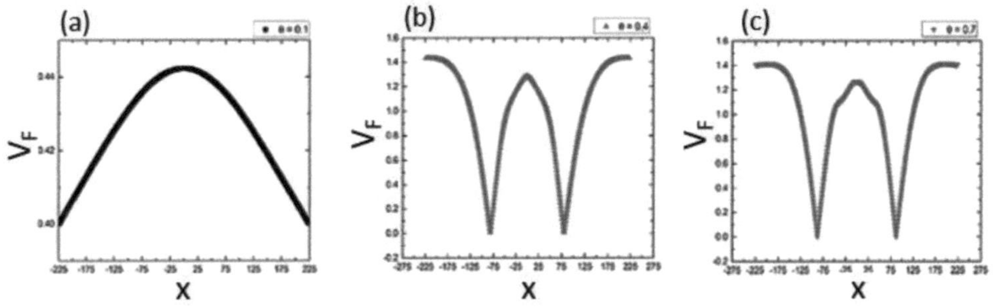

Figure 7.
Transversal cross section for $\xi = 0.3$ (a), $\xi = 1$ (b) and $\xi = 5$ (c).

different flow velocities as the distance between the maxima of the three structures does not remain constant during expansion. Supplementary investigations were performed by implementing similar data treatment for the Y direction. For the cross section on the Y axis (at X = 0) we can report a more fuzzy separation. This result can be explained as the structuring phenomena of the plasma is not limited to a unique flow axis, being observed in all directions. Moreover, the fractality of the our multifractal system, defined here through ξ and μ, is directly related to the trajectory of the plasma *particles*. As such, for a more complex plasma model (multi-element, multi-structured) the intrinsic dynamic within the plasma structures will induce to a separation on the main expansion axis (at X = 0).

This rather complex multifractal theoretical approach manages to simulate the structuring of a multielement (complex) plasma flow. Nevertheless, this is remains an abstracted view to a real dynamics in various technological application. For the validation of the conceptual and mathematical approach we chose to perfume comparisons with our experimental investigations of laser produced plasmas in quasi-identical conditions to those generally used for pulse laser deposition. Our model is suitable for the description of PLD physical phenomena as in past years various groups have shown [26–29] that in the case of multi-element plasmas there is axial and lateral segregation of the plasma particles during expansion based on their physical properties (mass, melting temperature), which damages the quality and properties of the deposited film.

5. Conclusion

The dynamics of a complex multi element plasma was investigated in the framework on a non-differential, multifractal theoretical model. Structuring into multiple plasma fragments were observed for the multi-fractal fluid like system containing structural units with various physical properties. The formation of complex plasma structures during expansion is corelated to the interaction between the transient plasma structural units and it is defined here by the complex phase of the velocity field and the fractalization of the particle geodesics. The multifractal system of equations was simplified by analyzing only two main directions. The plasma splits in multiple structures symmetrically to the main expansion axis.

The multifractal theoretical model was compared with empirical investigations of transient plasmas generated by laser ablation of a multielement metallic targets. The expansion of the plasma plume was monitored by means of ICCD fast camera photography and optical emission spectroscopy. The ICCD fast camera imaging showcased the formation of two or three main plasma structures in the main expansion direction, coupled with a similar phenomenon in the transversal direction. This complex behavior affects the angular plasma expansion and subsequently affect the spatial distribution of the deposited film. The heterogeneity of the plasma plume velocity field is in good agreement with the theoretical assumption presented in the framework of the non-differential model.

ICCD imaging revealed the splitting of the laser produced plasmas into two different structures, expanding with different velocities. An angular distribution of the front velocity was reconstructed for each of the two plasmas. The specie velocities were correlated to the properties of the elements found in the target (mass and conductivity).

A novel theoretical approach based on multifractal physics was used to simulate the behavior of multi element plasmas. The model considers the relation between the scattering probability, collision frequency and the fractality degree of the plasmas. The angular distribution of the ejected particles was discussed with respect to the fractality of the system. The simulation results are in good agreement with the experimental data.

Acknowledgements

This work was supported by Romanian Ministry of Education and Research, Nucleu Program LAPLAS VI – contract n. 16 N/2019 and contract n. PD-145/2020.

Conflict of interest

The authors declare no conflict of interest.

Author details

Stefan Andrei Irimiciuc[1*], Norina Forna[2], Andrei Agop[3], Maricel Agop[4], Stefan Toma[3] and Doriana Forna Agop[2]

1 National Institute for Laser, Plasma and Radiation Physics, 409 Atomistilor Street, 077125 Bucharest, Romania

2 "Gr.T.Popa" University of Medicine and Pharmacy – Iaşi,Str. Universităţii no. 16, 700115, Iaşi, Romania

3 Material Science and Engineering Department, "Gheorghe Asachi" Technical University of Iasi Romania, 700050, Iasi, Romania

4 Department of Physics, "Gh. Asachi" Technical University of Iasi, 700050, Iasi, Romania

*Address all correspondence to: stefan.irimiciuc@inflpr.ro

References

[1] Ojeda-G-P, A., Schneider C.W., Döbeli M., Lippert T., Wokaun A. 2017. Plasma plume dynamics, rebound, and recoating of the ablation target in pulsed laser deposition. Journal of Applied Physics 121 (13): 135306.

[2] Vitiello, M., Amoruso S., Altucci C., de Lisio C., Wang X. 2005. The emission of atoms and nanoparticles during femtosecond laser ablation of gold. Applied Surface Science 248 (1–4): 163–166.

[3] Santos A., Lopes Barsanelli P., Pereira F. M. V., and Pereira-Filho E. R. 2017. Calibration strategies for the direct determination of Ca, K, and Mg in commercial samples of powdered Milk and solid dietary supplements using laser-induced breakdown spectroscopy (LIBS). Food Res. Intern. 94: 72–78.

[4] Diwakar, P. K., Harilal S. S., Hassanein A., Phillips M. C. 2014. Expansion dynamics of ultrafast laser produced plasmas in the presence of ambient argon. Journal of Applied Physics 116 (13): 133301.

[5] Zhou, L. 2017. Two-fluid turbulence modeling of swirling gas-particle flows — A review. Powder Tech. 314: 253–263.

[6] Qiu, Y., Deng B., Kim C. N. 2012. Numerical study of the flow field and separation efficiency of a divergent cyclone. Powder Techn. 217: 231–237.

[7] Irimiciuc, S. A., Bulai G., Gurlui S., Agop M. 2018. On the separation of particle flow during pulse laser deposition of heterogeneous materials - a multi-fractal approach. Powder Tech. 339: 273–280.

[8] Nedeff, V., Lazar G., Agop M., Eva L., Ochiuz L., Dimitriu D., Vrajitoriu L., C. Popa. 2015. Solid components separation from heterogeneous mixtures through turbulence control. Powder Techn. 284: 170–186.

[9] Nedeff, V., Lazar G., Agop M., Mosnegutu E., Ristea M., Ochiuz L., Eva L., Popa C. 2015. Non-linear Behaviours in complex fluid dynamics via non-differentiability. Separation Control of the Solid Components from Heterogeneous Mixtures. Powder Techn. 269: 452–460.

[10] Kelessidis, V. C., and G. Mpandelis. 2004. Measurements and prediction of terminal velocity of solid spheres falling through stagnant Pseudoplastic liquids. Powder Tech. 147 (1–3): 117–125.

[11] Zhang, S, Kuwabara S., Suzuki T., Kawano Y., Morita K., Fukuda K. 2009. Simulation of solid-fluid mixture flow using moving particle methods. J. Com. Phys. 228 (7): 2552–2565.

[12] Monaghan, J.J. 1992. Smoothed particle hydrodynamics. An. Rev. Astro. Astrophys., 543–74.

[13] Irimiciuc, S., Bulai G., Agop M., Gurlui S. 2018. Influence of laser-produced plasma parameters on the deposition process: In situ space- and time-resolved optical emission spectroscopy and fractal modeling approach. Appl. Phys. A: Mat. Sci. Process. 124(9):615

[14] Irimiciuc, S.A., Gurlui S., Nica P., Focsa C., Agop M. 2017c. A compact non-differential approach for modeling laser ablation plasma dynamics. Journal of Applied Physics 121 (8).

[15] Ngom, B. D., S. Lafane, S. Abdelli-Messaci, T. Kerdja, and M. Maaza. 2016. Laser-produced Sm1−xNdxNiO3 plasma dynamic through Langmuir probe and ICCD imaging combined analysis. Appl. Phys. A: Mat. Sci. Proces. 122 (1): 1–7.

[16] Singh, J., R. Kumar, Awasthi S., Singh V., Rai A. K. 2017. Laser induced breakdown spectroscopy: A rapid tool for the identification and quantification of minerals in cucurbit seeds. Food Chemistry 221: 1778–1783.

[17] Irimiciuc, S. A., Mihaila I., Agop M. 2014. Experimental and theoretical aspects of a laser produced plasma. Physics of Plasmas 21: 093509.

[18] Anoop, K. K., Polek M. P., Bruzzese R., Amoruso S., Harilal S. S. 2015. Multidiagnostic analysis of ion dynamics in ultrafast laser ablation of metals over a large Fluence range. Journal of Applied Physics 117 (8).

[19] Geohegan, D. B. 1992. Fast-Iccd photography and gated photon counting measurements of blackbody emission from particulates generated in the KrF-laser ablation of BN and YBCO. MRS Proceedings 285 (January): 27.

[20] Geohegan, D. B., Puretzky A. A., Duscher G., Pennycook S. J. 1998. Time-resolved imaging of gas phase nanoparticle synthesis by laser ablation. Applied Physics Letters 72 (23): 2987–2989.

[21] Merches, I., Agop M.. 2015. Differentiability and Fractality in Dynamics of Physical Systems. World Scientific.

[22] Agop M., Merches I. 2019, Operational Procedures Describing Physical Systems, CRC Press, Florida.

[23] Cristescu, C.P. (2008).Non-linear Dynamics and Chaos : Theoretical Fundaments and Applications. Bucharest: Romanian Academy Publishing House.

[24] Mandelbrot, B. (2006).The Fractal Geometry of Nature. New York: W.H. Freeman And Company.

[25] Nottale, L. (2011).Scale Relativity and Fractal Space-Time : A New Approach to Unifying Relativity and Quantum Mechanics. London: Imperial College Press.

[26] Canulescu, S., E. L. Papadopoulou, D. Anglos, Th. Lippert, C. W. Schneider, and A. Wokaun. 2009b. Mechanisms of the laser plume expansion during the ablation of LiMn2O4. J.Applied Physics 105 (6): 063107.

[27] Canulescu, St., Döbeli M., Yao X., Lippert T., Amoruso S., Schou J. 2017. Nonstoichiometric transfer during laser ablation of metal alloys. Phys. Rev. Mat. 1 (7): 073402.

[28] Sloyan, Katherine a., Timothy C. May-Smith, Robert W. Eason, and James G. Lunney. 2009. The effect of relative plasma plume delay on the properties of complex oxide films grown by multi-laser, multi-target combinatorial pulsed laser deposition. Appl. Surf. Sci. 255 (22): 9066–9070

[29] O'Mahony, D, Lunney J., Dumont T., Canulescu S., Lippert T., Wokaun A. 2007. Laser-produced plasma ion characteristics in laser ablation of Lithium Manganate. Appl. Surf. Sci. Science 254 (4): 811–815.

Chapter 8

Laser Chemical Elemental Analysis: From Total to Images

Renata S. Amais, Danielle S. Francischini, Pedro S. Moreau and Marco A.Z. Arruda

Abstract

This book chapter focuses on laser ablation employed in elemental analysis and discusses the fundamentals and instrumentation of the laser-induced breakdown spectroscopy (LIBS) and laser ablation inductively coupled plasma mass spectrometry (LA-ICP-MS) techniques. The analytical performance of such techniques, challenges related to calibration, and strategies to improve sensitivity are discussed. In addition, the processes involved in data acquisition and imaging for acquiring the elemental spatial distribution are highlighted, and some representative examples in environmental, biological, medical, and forensic researches are presented.

Keywords: LIBS, LA-ICP-MS, imaging, plasma, mass spectrometry, optical emission, laser ablation

1. Introduction

The association of chemistry and light is so old than human being history, most probably raging from the observation of the solar spectrum to spectroscopic analysis, thus allowing the discovery of new and unknown substances. The advent of laser (acronym to Light Amplification by Stimulated Emission of Radiation) has only served to strengthen this natural bond. With the advances in electronics and computational programs, the lasers have a multitude of applications, from proteins (*i.e.* matrix-assisted laser desorption/ionization, MALDI) [1] to elemental analysis (*i.e.* laser ablation inductively coupled plasma mass spectrometry, LA-ICP-MS, or laser-induced breakdown spectroscopy, LIBS) [2]. The applications are dictated by the laser wavelengths, which, in fact, reflect their energies. For example, for biomolecules, like proteins, wavelengths higher than 350 nm are currently used, and for elemental analysis, 213 or 193 nm are most common.

Because of its modulated power, directionality, and temporal coherence, the laser has become a highly versatile tool, and used in a large variety of applications, from the study of how chemical reactions occur, then to initiate chemical reactions upon irradiation to extremely sensitive and selective means to evaluate the presence of chemical substances of interest [3]. Additionally, and as highlighted by the adage "a picture is worth a thousand words" [4], the laser ablation imaging [5] is another excellent option to greatly enhance the understanding of a studied system.

Then, this Chapter is devoted to elemental analysis, focusing on not only laser ablation ICP-MS and LIBS for qualitative and quantitative analysis, but also on imaging the results obtained from those analyses. Particular emphasis is placed on the discussion regarding the instrumentation, some processes involved in the image acquisition and formation, and also on the analytical results and figures of merit for a diversity of methods involving different areas as geochemical, environmental, biological, medical, and forensic. Some trends aspects and perspectives in the application of laser in chemical analysis are also the focus of this Chapter.

2. Laser-induced breakdown spectroscopy (LIBS)

The spectrochemical analysis exploits the electronic quantized transitions, which are characteristic of each individual element. An energy source (i.e. flame, plasma, laser, arc) can excite atomic species that emit specific wavelengths or frequencies upon returning to their fundamental state. The emitted light is spectrally resolved and detected to determine the elemental composition of the sample. The use of laser for the ablation process in the spectrochemical and elemental analysis was first proposed by Brech and Cross in 1962 [6]. At that time, the luminous plume produced by the ruby laser on the surface of metallic and non-metallic materials was able to remove a small mass of the target in the form of atoms and small particles, but too weak to provide usable spectra. Thus, an auxiliary electrical spark for excitation of vaporized and atomized material was added [7]. This study marks the conception of laser-induced breakdown spectroscopy (LIBS). Although the instrumentation proposed by Brech and Cross was soon commercially available, the interest in using LIBS rise in the 1980s. Advances in technology, more specifically the development of powerful and robust lasers, high-resolution optics, high sensitive detectors, and fast electronics for data acquisition have all contributed to its acceptance in routine analysis and increasing adoption as a cost-effective alternative analytical technique. Besides the ablation of the sample surface, using powerful modern lasers a microplasma is formed that excites the ablated atoms, ions, and molecular fragments. The plasma continues this excitation, and can also vaporize small ablated particles, atomize and excite atoms, ions, and molecules obtaining a rich emission spectrum used for qualitative and quantitative analysis [8]. LIBS is a highly versatile and adaptable spectroscopy technique. It has increasingly become a powerful tool for fast multi-elemental analysis in several fields of applications such as industrial, agriculture, environmental, food, geological, and biomedical. The growing interest on LIBS is probably due to the simplicity of the technique and instrumentation as sampling and subsequent excitation of atoms, ions and molecules rapidly occurs in one step and in the same system, i.e. there is no transportation of ablated material to another instrument as in LA-ICP-MS. It is a universal technique because any type of sample can yield a LIBS spectrum [8, 9]. The most notorious advantages of LIBS are the direct solid analysis capability, the quasi non-destructive analysis, and the portable instruments for analysis in the field. On the other hand, the technique presents limited detectability and it is not useful for trace element analysis unless physical and chemical enhancement strategies are applied to overcome this drawback.

2.1 Fundamentals and principals of operation

In practice, LIBS is a very simple spectroscopic technique to implement. A high-powered density pulsed laser beam is focused through lenses in (liquid or gaseous) or on (solid) the sample to produce dielectric breakdown leading to plasma formation. Plasma is a partially ionized gas containing atoms, ions and free electrons, and electrically neutral. Once initiated, the plasma induces the ablation of a finite quantity of the sample surface to a condition that may then be excited by the energy supplied by the same pulse or by a subsequent pulse of the laser beam (double pulse strategy) [9, 10]. Excitation is followed by emission of electromagnetic radiation which is collected, often through a fiber optic cable, and directed into a spectrometer where it is spectrally resolved and further instrumentally detected.

The interaction between a laser pulse and mater to create LIBS plasmas involves a process dependent on characteristics of both the laser, *e.g.* wavelength, irradiance and pulse duration, and the sample, *e.g.* gases, liquids, and solids, which can be conductive and non-conductive samples. When the laser pulse in the nanosecond time regime reaches the sample, the dominant mechanism is the thermal ionization process [9, 10]. A process of scattering transfers the laser energy to the lattice of the spot targeted on the sample causing the melting and generation of larger particles, thus enlarging the size distribution of the aerosol particles [11]. Plasma life stages include plasma ignition, plasma expansion and cooling and particle ejection and condensation. The physics of the plasma generation, evolution, and termination processes are complex. In the case of the analysis of solid samples, *i.e.* the majority of LIBS applications, plasma is initiated when the power density of the laser exceeds the breaking limit of a solid surface. In general, irradiance above 10^8 W cm^{-2} is needed. The breakdown is promoted by the intense electric field gradient of the laser, and atoms, ions, and electrons result from the deposition of energy into the target. Afterward, a high-pressure vapor is produced while the plume compresses the surrounding gas. A shockwave at supersonic speed is generated from the surface towards the surrounding atmosphere during vapor expansion. The absorption of the incident laser energy and transfer to the plasma occurs through the inverse Bremsstrahlung absorption involving interactions between free electrons, atoms, and ions. Free electrons in the hot vapor absorb photons from the incident laser increasing their kinetic energy to ionize additional atoms by collisions. The new electrons absorb more photons from the remaining pulse so that a cascade of ionization is generated [8–10]. Plasma is created having distinctive characteristics, high temperatures (10.000 K), and high electron densities ($>10^{17}$ cm^{-3}). The removed material in the ablation process of the solid samples contributes to plasma formation and expansion. In fact, the plasma is only sustained due to the presence of the ablated material [9]. The plasma expands in a similar way the initial breakdown occurs, at supersonic velocities producing a shockwave, also through the process of inverse Bremsstrahlung absorption. Meanwhile, electron number density and temperature of the plasma changes. The remaining laser energy from the laser duration of the pulse is continuously absorbed by the plasma. Collisions with the surrounding gas reduce plasma velocity of propagation, and then plasma cools down by self-absorption and recombination between electrons and ions, generating neutral species and clusters after plasma extinction [8–10]. Part of ablated mass are particles and these particles create condensed vapor, liquid sample ejection, and solid sample exfoliation, not interesting for LIBS analysis [10].

After 10^{-9} s and 10^{-8} s of the ignition, plasma induced by laser in the nanosecond regime becomes opaque due to laser pulse absorption by electrons in the plasma. It reduces the ablation rate because only a fraction of radiation reaches the sample surface. This phenomenon is called shielding and induces a crater with melted and deposited material around it [10]. Meanwhile, plasma is reheated, and the lifetime and size of plasma are higher. On the other hand, ultra-short pulses (femtoseconds) lead minimum plasma shielding and crater with highly defined edges without melted or deposited materials [9, 10]. This regime of pulse is too short to induce thermal effects in the breakdown process. Electronic excitation and ionization, and Coulomb explosion, are the main bond breaking and plasma ignition mechanisms, making the ablation mechanism mostly based on photochemical reactions resulting in vaporization presenting sharper size-distribution solid aerosol than those generated by nanosecond lasers [11]. In the case of picosecond laser pulses, both thermal and nonthermal processes can occur depending on the laser irradiance.

Initially, the emission spectrum is dominated by a background continuum, while ionic and atomic emission increases with time. The continuum is the main source of background signal (BG) in LIBS, which predominates at the first instants of the plasma life [12]. It is a result of the bremsstrahlung emission (radiant loss of energy due to electron deceleration) and banding (e.g. OH, N_2^+, NH, and NO). A delay time is required to start analytical measurement to avoid high continuous emission. The decrease in the emission intensity of the continuum occurs at a higher rate than the excited atoms or ions in the plasma. Thus, temporal separation is feasible and detectability is improved. In LIBS analysis, the delay time must be experimentally evaluated in order to obtain maximum signal-to-background ratio [12].

Undesirable matrix effects may lead to inaccurate determinations and lower sensitivity. These are usually a consequence of i) physical properties of the sample which change the ablation parameters altering the amount of ablated mass, ii) the presence of an element alters the emission features of another one, and *iii)* the plasma–particle interaction processes, which are time and space-dependent due to the transient nature of the plasma and its spatial inhomogeneity [13]. Elemental fractionation is defined as a non-stoichiometric effect which also depreciates the quality of the results [10, 13]. It occurs when the ablated material failure to represent the real composition of a sample due to preferential evaporation of volatile elements, selective segregation, surface temperature distribution inhomogeneity, among others. Elemental fractionation and matrix effects in laser sampling-based spectrometry methods have been discussed in detail by Zhang et al. [14]. Adequate laser wavelength, energy density, pulse width, and proper calibration using standard reference materials with known composition and matrix-matching strategy can overcome these drawbacks. An advantage of LIBS compared to LA-ICP-MS is fractionation can occur only during ablation, as there is no transportation of ablated material from ablation chamber to excitation/ionization source, or from ICP to mass spectrometer.

Plasma optical thickness is an important parameter in laser-induced plasmas. The emitted radiation is successively reabsorbed by atoms/ions located in the coldest plasma region leading to self-absorption and pronounced non-linear effects [15]. The plasma is called optically thick and usually occurs for the most intense emission lines of elements and for less intense emission lines at higher elemental concentrations.

2.2 Instrumentation

The main components of a generalized LIBS apparatus include (i) laser source; (ii) focusing lenses; (iii) sample support; (iv) optical fiber; (v) spectrometer (vi) detector and (vii) computer for precise control of temporal events, such as: pulse trigger laser and spectrum recording. In the case of a particular application, the specification of each component may be considered and changed. Elements to be monitored, expected concentration, type of analysis (quali or quantitative), sample characteristic (physical state, homogeneity and matrix composition) are common factors to consider when selecting the instrumental specifications.

The laser source generates the laser beam, which main properties are the wavelength and pulse width, both dependent on the laser source and its technological developments. The initial works employing laser in the chemical analysis used visible and infrared (IR) laser sources, such as 693 nm ruby [7, 16] and 1064 nm Nd:YAG (neodymium-doped yttrium aluminum garnet) [17] with a pulse width of μs and ns, respectively. However, further researches demonstrated that the ablation efficiency of transparent samples using visible or IR wavelengths was harmed due to the poor absorbance of the laser energy by these types of samples.

The evolution of optical technologies led the development of Nd:YAG with wavelengths of 266 and 213 nm by quadrupling and quintupling, respectively, the natural frequency of the Nd:YAG emitted laser. On the other hand, the use of gas excimer as a laser source also enhanced the possibilities of shorter laser wavelengths, such as 193 and 157 nm ArF lasers, which works in the deep-UV region [8–10]. As previously mentioned, the pulse duration is also an important parameter of the laser beam and impacts on the mechanism of interaction between the laser beam and the solid target, the amount of the ablated material, and the crater shape [8–10]. Nowadays, the typical pulsed laser used in LIBS works with a pulse width of nanoseconds or femtoseconds. The best type of laser used for LIBS depends on the application and the desired laser wavelength and pulse duration.

Laser pulses can be focused on the sample using lenses or mirrors. Focal length, diameter, and material are important parameters to achieve minimum spot size (highest power density on target), maximum transmission, and minimum back reflections. For systems requiring an adjustable focus, i.e. lens-to-sample distance may change, a multi-lens system may be required [17]. The collection of the emitted radiation and direct it into the spectrometer is possible by employing lenses and fiber optic or only lenses. Fiber optic transmits the light using total internal reflection, and it is especially useful when the detection system cannot be positioned close to the sample target [17]. A combination of lens and fiber optic is typical as the lens collimates the emitted light improving the focalization into the fiber probe [10].

Once emitted light by the plasma reaches the spectrometer, it is diffracted in order to obtain spectra in terms of signal intensity as a function of wavelength. Czerny Turner is a sequential dispersive system that combines two collimating mirrors and one grating, while Paschen-Runge optics use a concave grating to separate wavelengths allowing simultaneous multielement analysis. Although both designs have been employed in LIBS for many years, Czerny Turner optical mounting is limited by the monoelemental characteristic in which sequential multielemental analysis is impossible in the case of inhomogeneous samples as the elemental composition varies shot to shot, and Paschen-Runge presents relatively low resolution [9]. High-resolution and high spectral coverage devices are essential in LIBS due to the complexity of the emission spectra. Echelle optics combine a low-density grating with a prism and is called an order-sorting device, i.e. light is diffracted in two dimensions. It offers typical spectral bandwidth of 5 pm and high spectral coverage emission, approximately from 200 to 900 nm, which can be recorded with a single laser pulse. In recent years, the Echelle spectrograph has been more extensively used, but require a two-dimensional detector [10, 17].

Detectors devices convert the diffracted optical signal into an electric signal. Photodiodes multiply the current produced by incident light striking photocathode by multiple dynode stages. This is the simplest and inexpensive device but useful only for one-dimensional spatial information from the spectrometer (e.g. Czerny Turner and Paschen-Runge). On the other hand, charge-coupled device (CCD) is constituted by a bi-dimensional configuration of several sensors, each made up of three electrodes over a common substrate of p-type silicon, which allow the acquisition of two-dimensional spatial information obtained by the Echelle system. More recently proposed, intensified CCD (ICCD) is a CCD coupled to microchannel plates to provide time-gated detection of the laser plasma [8–10]. Thus, time-resolved detection down to a few nanoseconds is possible, which is essential to avoid high continuous emission [17].

2.3 Analytical features and strategies to improve sensitivity

The majority of LIBS measurements involve the analysis of solids. Some applications are limited by the relatively low sensitivity of the technique. Limits of detection of LIBS usually range from 1 to 100 parts per million (mg kg^{-1}). Therefore, most applications are focus on major elements, as the technique cannot meet the demands for the detection of trace elemental analysis (parts per billion, μg kg^{-1}). Physical and chemical strategies have been demonstrated to enhance the LIBS detection limits and sensitivity [18]. Increased plasma temperature and electron density are achieved by double-pulse laser method using two laser sources, the use of spatial and magnetic constraint devices, and controlling the atmosphere in which the sample is placed with inert gas (e.g., N_2, Ar, and He) [19]. Nanoparticles (NPs) deposited on surfaces of the solid samples favors ablation processes which mechanisms differ for conductors and insulators samples. For liquid sample analysis, liquid–liquid extraction, liquid–solid conversion, and surface-enhanced LIBS (liquid sample is dried onto the surface of a selected solid substrate before the analysis) have been exploited to overcome problems due to laser-liquid sample interaction, laser energy dissipation, low plasma temperature and sample splashing which depreciate repeatability and reproducibility. The reader is referred to a recent review paper by Fu et al. [18].

The possibility of measuring the molecular emission in LIBS allows the determination of some non-metallic elements from emission bands of diatomic molecules, e.g. fluorine and chlorine have been detected by the emission of CaF, BaF, MgF, CaCl, SrCl, or MgCl [20, 21]. Isotopic analysis based on the discrimination between emission bands of molecules formed by two different isotopes has been reported. The different masses of the isotopes affect the vibrational and rotational energy levels results in molecular isotopic shifts which are exploited in the isotopic determinations [21].

Different data acquisition modes are possible in LIBS analysis and it is selected depending on the goal of the experiment. Using just one laser pulse or using repetitive laser pulses, localized microanalysis with lateral and depth profiling information is easily obtained. For image-based analysis, a generation of a series of plasmas at different positions on the sample following a scan sequence is necessary. Most of LIBS imaging instruments rely on an XY stage that moves the sample instead of moving the laser beam because of a greater collection efficiency from a fixed plasma plume. LIBS imaging analysis is later discussed in this text, and additional information can be found in the review by Jolivet et al. [22].

An important innovation in LIBS is the handheld instruments commercially available for analysis in the field, especially useful when the sample cannot be moved. Some instruments present capability of chemometric analysis by means of proprietary software, video targeting, and an argon purge of atmosphere neighboring the target in order to improve sensitivity. Applications in agriculture, environment, industry, and cultural heritage can provide information to solve important economic and historical issues. An impressive breakthrough of this technique is the use in planetary geology on a Mars mission for remote sensing.

3. Laser ablation inductively coupled plasma mass spectrometry: LA-ICP-MS

With the development of laser technologies and the increasing demand for specific applications in the direct analysis of solids, in 1985 Alan Gray [16] demonstrated the coupling between a ruby laser ablation system with an inductively coupled plasma mass spectrometer (LA-ICP-MS) for monitoring Si, Al, Fe, Ca, Mg, Na, K, Mn, and Ti in pelletized standard powdered rock samples. Since its application, the area has been growing and one indicator of that is the number of scientific works published: more than 15,000 research papers related to LA-ICP-MS in the last 6 years (source: www.sciencedirect.com, September 10th, 2020).

The interest in the LA-ICP-MS technique is related to its impressive characteristics, such as the possibility to acquire elemental and isotopic information by direct sampling of nano, and even pico or femtograms of solid materials in the micrometric scale, and limits of detection around low parts per million (in some specific cases hundreds of parts per billion). Additionally, the image-based analysis to study sample heterogeneity and elemental or isotopic distribution in a surface contributed to its acceptance [23, 24]. These almost unique characteristics permit the use of LA-ICP-MS in many fields of science, such as geochemistry, forensic, environmental, materials, medical, and biological, in specific applications that were not possible by conventional elemental analysis and sample pretreatment.

3.1 Principles of operation

The LA-ICP-MS technique works based on the coupling of a solid sampling system, the LA system, with a powerful elemental/isotopic analytical technique, the ICP-MS. A high-power laser beam, focused on the sample surface through an optical system promotes a huge and instant increase of the temperature of the target sample and transfers a discrete sample volume to the vapor phase. A solid aerosol is generated in an ablation chamber because of this interaction between the laser beam and the sample, and this process is named laser ablation [2, 25].

The solid aerosol generated carries the information of the elemental and isotopic composition of the sample ablated (analytes and matrix components). To acquire this information in terms of analytical data, the aerosol is transferred to the ICP-MS by a gas stream through a connection tube to the ICP. In the ICP, a high-temperature argon plasma induced by a radiofrequency, the aerosol is digested, and its solid constituents are vaporized, atomized, and ionized [2, 25].

The ions generated into the ICP are extracted, through the interface region, to the mass spectrometer (MS), the instrumental component responsible to separate and detect these ions. After the extraction by the interface region, the ion optics conduces the ion beam up to the mass analyzer device, a high vacuum region in which separation of the ions based on its mass-to-charge ratio (m/z) occurs. The ions are further detected by a specific detector that generates an electric impulse due to the interaction of the ions on its surface [2, 25].

The analytical signal profile observed in a typical LA-ICP-MS analysis is a transient signal characterized, in general, by the counts per second (intensity) of the monitored ion as a function of the ablation time. The specific characteristics of the LA-ICP-MS instruments available for chemical analysis are dependent on each instrumental component type and its principles of operation discussed in the next sub-section.

3.2 Instrumentation and its fundamental characteristics

As mentioned before, the LA-ICP-MS permits different instrumental setups that can impact on mechanisms of the interaction between the laser beam and the solid sample, the separation of the ions in the mass analyzer, and the analytical response of the instrument during analysis. The main components of LA-ICP-MS instrument are the laser source, ablation chamber equipped with a CCD camera, and transport tubing connected to the ICP-MS. The ICP-MS, in turn, can be composed of different mass analyzer types, which presents specific analytical features, and the detector. The main characteristics of each LA-ICP-MS component are discussed below, focusing on principles of operation, commenting on the potentialities and limitations, as well as its technological developments and improvements.

3.2.1 Laser ablation system

The laser source is the heart of a laser ablation system because it confers the main properties of the laser beam and thus, the main characteristic of the ablated material. It was previously mentioned, in the LIBS section, the coupling between laser radiation and the solid sample surface promotes the ablation depending mainly on the laser wavelength and pulse width. Readers are referred to as sections 2.2 and 2.3. Besides the efficient coupling, these parameters impact the size-distribution of the aerosol particles and the crater shape. In general, the ablation rate increases by increasing the laser energy (shorter wavelengths), and smaller particles of the solid aerosol are obtained using femtosecond lasers, which are required to guarantee the efficiency of the transport and posterior digestion of these particles into the ICP. The most common pulsed laser used in LA-ICP-MS present nanoseconds or femtoseconds pulse width and short wavelengths (e.g. 213 nm, 5th harmonic solid-state Nd:YAG laser).

Different from LIBS which sample holder can be open support, in LA coupled systems, such as LA-ICP-MS, a closed gas-tight compartment denominated ablation chamber is required. The ablation process occurs in the ablation chamber, which presents a transparent window in the wavelength of the laser beam, and the gas streaming transports the aerosol to the ICP. The ablation chamber is coupled to a CCD camera to improve sample visualization and to define the specific position of the sampling target on the sample surface using the LA software. The types of ablation cell designs are vast, including commercial and customized, but some requirements are needed for the efficiency of the role of ablation chambers.

In general, ablation chambers must provide a gas environment that permits an expansion of the aerosol generated to ensure a small-size particle-distribution, and an adequate gas-flow for an efficient transport of the solid aerosol to the ICP avoiding memory and fractionation effects. Thus, the ablation volume, the flow dynamics of the gas stream, and the diameter and length of the transport tube are determinant for the features of an efficient ablation chamber [26]. The carrier gas also plays an important role in chemical analysis. Lighter carrier gases, such as helium, improves the efficiency of the ablation process and particle transportation enhancing sensitivity of the analytical method and reducing drawbacks related to fractionation in the transport process [27]. Recent developments in the design of low dispersion systems, rapid response ablation cells and their impact on bioimaging applications was recently discussed by Van Malderen et al. [28].

3.2.2 ICP and mass analyzer devices

LA system generates a laser beam that is responsible for removing an amount of sample and transferring its constituents to the vapor phase. From this point, is the role of the ICP-MS to analyze this material and provide analytical information regarding its elemental and isotopic composition. Because the analytical entities monitored by ICP-MS are mainly mono-charged positive ions, ICP-MS instruments must first be able to generate ions from the aerosol constituents of the sample. The ICP's are ion sources in which plasmas are produced by the energy transferred from a radiofrequency generator to a gas flow into a concentric quartz tube (torch) via a magnetic field through an induction coil [29]. The plasma has the ability to ionize most atoms of the elements presented in the periodic table due to its high-temperature (around 10,000 K) and high electron density. The inductively coupled plasma formation mechanism will not be discussed in this chapter, but we encourage the readers to look for it in Thomas, 2013 [30].

As the ions of the sample composition are formed in the plasma, including the analytes and concomitants, they are conducted towards the interface region to enter the mass spectrometer. This interface region is composed of metallic cones, normally two (sampler and skimmer cones) which extracts these ions. Then, ions go to the mass spectrometer via the ion optics, which are a set of electrostatic lenses that conduce the positive ions to the mass analyzer, and deviates the neutral and negative species, as well as the photons [30].

The heart of the mass spectrometer is the mass analyzer. This component has the function to, based on the mass-to-charge ratio, separate the analyte ions from concomitants, permitting the detector count and acquire the abundancy of the specific element under analysis. For this, there are different arranges of mass analyzers with their principles of operation to separate the ions. The most employed mass analyzer types in LA-ICP-MS are quadrupole systems, magnetic sector, and time-of-flight [30].

The quadrupole system is composed of two pairs of metallic bars, positioned oppositely. It operates as a mass analyzer by applying specific direct and alternate current (AC/DC) in the pair of rods that in consequence, for each specific AC/DC applied, only the desirable m/z of the ions presents a stable trajectory in the quadrupole, reaching the detector. The trajectory of the other ions will be unstable and will exit the quadrupole before reaching the detector. This mass selection occurs sequentially, and each m/z determined has an optimum AC/DC voltage applied on the quadrupole. This characteristic, as well as the design of the quadrupole, confers a resolution ranging from 0.7–1.0 *a.m.u*. It is possible to optimize the parameters of quadrupole to increase the mass resolution, but a compromise with the sensitivity must be attempted. This relatively low resolution impacts some applications due to the interferences that can be present in the analysis. Some isobaric and polyatomic species formed in the plasma can present the same m/z of the analyte and then depreciate the accuracy of the analysis. For this, the new quadrupole-based ICP-MS's presents a collision/reaction cell or interface that uses a reactive or non-reactive gas to break polyatomic interference species, reduce their kinetic energy or react with the analyte generating a new analyte ion free of interference, then allowing accurate results. Although the limitations of single quadrupole devices in terms of resolution, the low price, and suitability for most quantitative applications make it the most widely used mass analyzer in ICP-MS analysis. The newest inductively coupled plasma tandem mass spectrometry, popularly called triple-quadrupole ICP-MS presents an additional quadrupole located before the collision/reaction cell. It confers the possibility to filter non-analyte ions in the first quadrupole, preventing the production of unwanted species that could immediately interfere on the m/z of interest and the second mass analyzer to deal with interferences using mass-shift or on-mass strategies. It has shown great results of accuracy in some specific difficult applications for single quadrupole ICP-MS [30].

Another mass analyzer type is the double-focusing magnetic sector technology. This mass analyzer device operates with two analyzers, a magnetic and an electrostatic. This design of mass analyzer presents greater resolution and interference overcoming compared to the quadrupole, and is applied in studies that need this resolution, such as isotopic ratio monitoring and analyte quantification in complex samples. The principles of m/z separation in such mass analyzer are based on the dependence of the deflection angle of the ion beam by a magnetic field to the mass of the ions, and the alignment of the ions with the same m/z in the electrostatic field, due to its kinetic energy. The combination of these two mechanisms of separation improves the resolution and the precision of the measurements. With a single detector, the mass separation is sequential, by varying the magnetic field in the deflection separation, allowing a specific m/z at a time to reach the outer slit and the detector. But, if this double-focused magnetic sector technology is arranged with an electrostatic sector field followed by a magnetic sector field and a detector plane, known as multi-collector, the detection of the ions is simultaneous and the precision of the ion measurement is improved. This technology is the state-of-the-art instrumentation for high precise and accurate isotopic ratio studies. The disadvantages of the double-focusing magnetic sector technology are related to the price of these instruments and, in the case of multi-collector, the loss of resolution, impacting on the analysis of complex matrices [30].

The last discussed mass analyzer technology is the time-of-flight (ToF) mass analyzer. Its principles of operation are based on the velocity dependence of the ions to the m/z, for ions generated at the same time and with the same kinetic energy. For this, packages of ions are simultaneously introduced in the mass analyzer by a specific ion optics, and, in a flight tube, the package of ions is accelerated due to a constant voltage, through a known distance. As the kinetic energy of the ions is the same, they will reach the detector in different times-of-flight due to their differences in m/z. In general, lighter isotopes reach the detector first, followed by medium m/z ions and, finally, heavier ions. These principles of operation can produce 20,000 mass spectra per second and, in contrast with the quadrupole technology, there is no dependence on the resolution and the sensitivity for the ToF analyzers. So, it came in the scientific community as an instrument with higher resolution compared with the quadrupole system, faster and is applicable not only for isotope ratio studies but also for fast qualitative screening and quantitative analysis of complex matrices [30].

As can be seen, there are many possibilities of LA-ICP-MS setup, whether the LA system, with the properties of the laser source and ablation chamber, or the ICP-MS specificities, such as the ICP conditions, mass analyzer devices, and their particularities, not mentioning the detector types, that are not less important, but were not discussed in this chapter due to de variety of designs in the market. In this way, LA-ICP-MS hardware's designs will provide specific analytical features for each instrument setup, that will be commented in general in the next sub-section, to a better understanding of LA-ICP-MS potentialities and limitations.

An innovative approach that exploits the combination of the two techniques, LIBS and LA-ICP-MS, is known as Tandem LA/LIBS and commercially available (**Figure 1**). It provides simultaneous and complementary information for total and spatially resolved mapping of major and trace elements. Emission spectra are simultaneously monitored from the micro plasma created by the laser during ablation/sampling.

3.3 Analytical features, challenges and limitations

The capabilities of LA-ICP-MS as an analytical technique starts with the characteristics of the LA system. It has the ability to direct sampling with micrometric spatial resolution, varying from 4 to 200 μm in most instruments. As the ablation chamber can move on the axis x and y, the ablation can be performed in spot analysis and also by making ablation lines over the sample area of interest, and this is the sampling principle for image-based LA-ICP-MS applications later discussed [2, 24, 25].

For quantitative applications, and especially using quadrupole-based ICP-MS, the LA-ICP-MS can achieve a precision of 2–5% for the determination

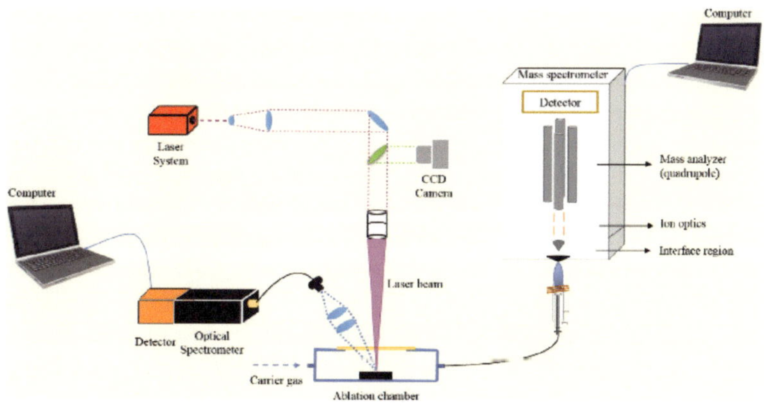

Figure 1.
Schematic diagram of the for simultaneous LIBS and LA-ICP-MS.

of homogeneous elements in solid samples. For isotopic ratio studies and using double-focusing magnetic sector ICP-MS with multi-collector, the LA-ICP-MS allows a precision, in some cases, of 0.001–0.005% in the isotope ratio measurement. For heterogeneous samples, the precision is not the goal of the analysis, but the evaluation of the elemental distribution and its concentration in different areas of the sample [2, 24, 25].

In terms of limits of detection, it varies depending on the sample matrix and instrumental setup employed, usually ranging from hundreds of ng g^{-1} to a few μg g^{-1}. Due to its limit of detection and the low mass ablated, LA-ICP-MS can cover a wide range of concentration, allowing the determination of trace, micro and major elements directly in the solid sample analyzed [2, 24, 25].

Although the potentiality of LA-ICP-MS, there are some drawbacks that the scientific community have devoted effort to overcome, and that the users must give special attention. The critical limitations of LA-ICP-MS are derived from the nature of the interaction between the laser beam and the solid sample and the transport process of the solid aerosol generated through the ICP. Since the mass removed by the laser beam is dependent on the matrix of the sample and the characteristics of LA system and parameters (laser fluence, spot size diameter, and repetition rate), thus, the analytical signal obtained in the ICP-MS is also matrix-dependent. This fact impacts the accuracy of the analytical method, which will be achieved in the condition that the same mass is ablated from the sample and the calibration standards, and the analyte undergoes the same transportation effects and processes in the plasma until be ionized [2, 25].

However, it is not simple to achieve a suitable solid calibration standard for the vast sample matrices that are analyzed by LA-ICP-MS. The ideal condition is the use of certified reference materials (CRMs), but a limited number of matrix types is available as CRM. It is especially difficult for the analysis of heterogeneous and non-powdered samples. To overcome this drawback, the scientific community has been studying different approaches to calibration. The main strategies are the use of matrix-matched materials, such as lab-made standards, by spiking the analytes in the powdered material that has approximately the same composition of the sample, drying and pressing it into a pellet. In the case of non-matrix-matched standards, solution-based strategies are employed, for example, the use of solution nebulization of liquid standard and mixing it with the aerosol of the sample, to calibrate the method. In all these cases, the homogeneity of the standard must be monitored to guarantee the analytical performance of the external calibration procedure. Additionally, internal standardization is usually needed to correct signal fluctuations during the analysis. A detailed review of calibration strategies for LA-ICP-MS quantification method was written by Miliszkiewicz et al. [31], and can be checked for more information about calibration strategies [2, 25].

Another drawback in LA-ICP-MS analysis is the fractionation that can occur in an ablation procedure or transport process. As previously mentioned in the LIBS section of this chapter, femtosecond UV lasers could suppress this effect, due to the minimization of the thermal process during the ablation. Aligned with the femtosecond UV laser, a well-designed ablation chamber and transport tubing will permit the small size distribution of the aerosol particles, avoiding fractionation during the ablation and the transportation through the ICP. Another point, and not less important, the optimization of the LA-ICP-MS instrument parameters are recommended to guarantee its suitability to the sample matrix analyzed. Such parameters are the laser frequency, spot size diameter, repetition rate, carrier gas flow rate, auxiliary gas flow rate, and radiofrequency power [2, 25].

In spite of these limitations, a careful dealing with LA-ICP-MS analysis, followed by an adequate optimization of the instrumental parameters, monitoring the possible sources of analyte fractionation, and a suitable calibration strategy allow the LA-ICP-MS users to achieve impressive information of solid samples that could not be accessed using traditional solution-based ICP-MS analysis, enlarging the possibilities of application.

3.4 LA-ICP-MS applications

The fields of science that LA-ICP-MS is applicable are vast, including forensic, environmental, materials, biological, geological, etc. Although the image-based analysis using LA-ICP-MS composes the state-of-the-art of this technique, the use of LA-ICP-MS in studies involving total elemental analysis are also useful and confers interesting information about plenty kinds of samples that are pointed out above.

The use of LA-ICP-MS in forensic science can provide information of crime evidence without destructing the sample (usually obtained in a small amount). For example, the elemental composition of tape packaging samples could be used to classify them according to their origin rolls [32], and the elemental composition of glass evidence also allows for forensic crime elucidation [33]. The analyses of difficult to prepare samples are feasible using LA-ICP-MS, which is the case of hair samples. Ash and He [34] demonstrated that LA-ICP-MS was used to understand the poisoning dynamics of thallium in a criminal case, including understanding the increasing of doses of poison and the time interval. Levels of Cu, Zn, and Hg monitored in hair samples of different grizzly bears showed adequate correlation with the duration of salmon consumption and the amount of it [35], which is an important environmental monitoring method of mammal's wildlife.

LA-ICP-MS is applied also in conjunction with separation techniques, such as thin-layer chromatography (TLC), to improve the quality of the data. The use of TLC-LA-ICP-MS could provide quantitative information of gold nanoparticles (AuNPs) by separating them from gold ions, as well as AuNP size information by using a specific mobile phase in the TLC separation [36]. TLC was also employed in fractionation studies of S, Ni, and V in petroleum using LA-sector field ICP-MS [37].

The reader is referred to critical reviews of recent LA-ICP-MS applications, such as by Lobo et al. [38], where isotopic analysis in biological studies are discussed; or Limbeck et al. [39], where the challenges and advances in the quantitative analysis are detailed. Another interesting work by Pozebon et al. [40] demonstrates the use of LA-ICP-MS analysis of biological samples bringing information of novel developments in instrumentation, methods of calibration, and applications in modern demands, such as single-cell analysis and NP uptake.

4. Elemental distribution imaging via LA-ICP-MS and LIBS techniques

Images are present in daily life of the population in different contexts, such as through the image created by the eyes, enabling observation of the environmental around us; through pictures and photographs that record different personal or historical moments; and also, via magnetic resonance imaging (MRI) or X-ray computed tomography (CT) which contribute to medical diagnostics. In the chemistry context, imaging is a process that transforms the spectral information of atoms and molecules present in the solid sample surface, in a high resolution image through the application of powerful spectral techniques as LA-ICP-MS, LIBS, Raman [41], X-Ray Spectroscopy [42] and Secondary Ion Mass Spectrometry (SIMS) [43].

For the LA-ICP-MS and LIBS techniques, the imaging process occurs through the applications of laser pulses directed at specific regions of the sample surface (*x,y* coordinate), via point by point or continuous lines performing the chemical measurements and obtaining spectral information of the region of interest [22, 44]. In this case, a document (.txt or .log format, for example) is generated for each point or a line containing the signal intensity data of all the elements measured [45]. Due to the complexity and a large number of data obtained, appropriate software for data processing is required to separate the information and generate a data matrix, thus allowing the generation of a final image for each element. A two (2D) or a three (3D) dimensions image can be created for each element in the sample. The 2D image is equivalent just to the surface image and the signal intensity or concentration of the analyte, obtained by the conventional method describe previously. The 3D image can be obtained in two ways, i) analyzing and combining each layer of the sample (volume reconstruction of several 2D images), ii) repeated pulses of laser in the same region, allowing in-depth elemental imaging [46]. The scheme of the imaging process *via* LA-ICP-MS or LIBS can be seen in **Figure 2**.

There are different software for data processing and generation of chemical images, such as Microsoft Excel, LA-iMageS [47], and MATLAB [48]. In the case of LIBS imaging, the use of chemometrics tools has been a great ally in the data treatment due to the complexity of the emission spectra data. Principal component analysis (PCA) and partial least squares (PLS) are commonly used, and also can

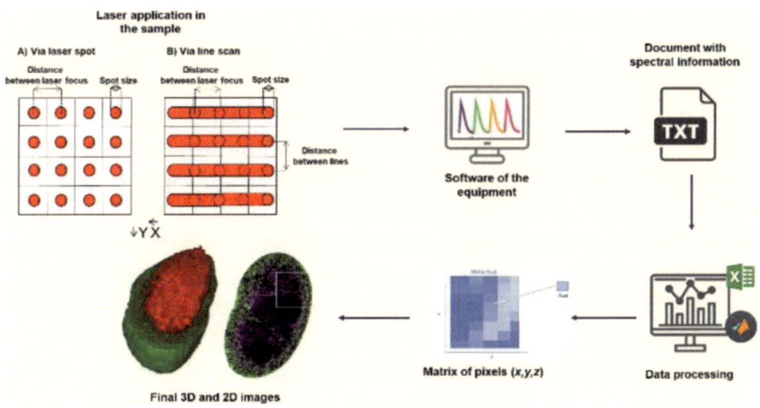

Figure 2.
Steps of the imaging process via LA-ICP-MS and LIBS. Final 3D and 2D nanoparticles images by Gimenez et al. [46], reproduced with permission of Springer Nature.

generate hyperspectral images [22, 49]. The final images are formed by a matrix of pixels, represented by I(x,y,z), in which x,y are the sample coordinates and z, the spectral information. Therefore, the number of pixels that will form the final image will be directly related to the amount of information, image quality, and the spatial resolution that the imaging process can provide.

The spatial resolution is the ability of the imaging method to distinguish two points in the sample surface, also mentioned as lateral resolution. It is correlated to the spot size and line scan direction, the x- and y-resolution. The x-resolution (μm) is related to the scan direction and the pixel size of each data set, obtained by multiplication of the scan speed and the acquisition time. The y-resolution (μm) is correlated to the distance between the lines. Thus, the spatial resolution and, consequently, the pixel size, is a result of the laser spot size, scan speed, acquisition time, and the distance between the measured lines or points. These parameters vary according to the application and optimized instrumental parameters. It is important to mention that LA-ICP-MS and LIBS are considered high spatial resolution techniques [48, 50, 51].

The imaging process has the advantage of accessing spatial and distribution information of chemical species that would not be possible using the conventional method of analysis, as atomic emission spectroscopy (AES) and inductively coupled plasma mass spectrometry (ICP-MS). These techniques usually require sample preparation steps, as decomposition and/or extraction, that cause destruction of the sample and, consequently, of spatial information. One of the major limitations of imaging via LA-ICP-MS and LIBS is the long data acquisition time, varying according to the size of the analyzed sample, scan speed, and laser parameters, as well as making the whole system result in representative images of the system in evaluation [22, 44]. Quantitative imaging based on LIBS and LA-ICP-MS is still a challenge and a controversial topic between researchers. Most methodologies are considered just qualitative or semi-quantitative, but in the literature, it can be observed different studies that aim to developed calibrations and mathematical strategies to transform the relative intensity information into a quantitative image [46, 52]. The applicability of the LA-ICP-MS and LIBS imaging can provide spatial chemical information to solve problems in different areas of science as demonstrated in the next section.

4.1 Applications of elemental distribution imaging via LA-ICP-MS and LIBS

Biomedical researches and clinical diagnosis are some of the most important areas of chemical imaging applications because is necessary to know the chemical specie involved in different diseases and the specific localization in human tissues, thus allowing specific drug development. Recent reviews of LIBS [22] and LA-ICP-MS [53] imaging applied in medical science can be found in the literature.

Moncayo *et al.* applied LIBS imaging of human paraffin-embedded skin samples as complementary biopsy with conventional histopathology sample preparation to detect the differences of the spatial distribution of P, Mg, Na, Ca, Zn and Fe in healthy and malignant skin tissues of cutaneous metastasis of melanoma, Merkell-cell carcinoma (MCC) and squamous cell carcinoma (SCC). In this study, two lens-fibred system and two spectrometers were used to realize the simultaneous detection of all elements. The high-resolution images were obtained using laser shots with 50 μm of step size and an acquisition time of 3 hours for each sample with, approximately, 3cm^2 of the tumor area. As imaging results, high levels of P were observed in all skins in the tumors areas; high levels of Ca and Zn were noticed closest to the tumor and decrease in the areas away from the tumor in the MCC; a gradient level of Ca and Mg in metastatic melanoma and Na, Mg and Zn in

SCC were observed in the left area of the tumor, allowed proper identification and discrimination of the three different skins analyzed by LIBS imaging [54].

Applying quantitative LA-ICP-MS imaging using line scan mode, the spot diameter of 10 μm, 20 μm s^{-1} scan speed, and matrix-matched calibration, Crone *et al.* could evaluate the ineffectiveness of Pt drug-based in bone metastasis treatment in a mouse tibia samples, in which are generally considered an unsatisfactory and scarce treatment. According to the results, there is no efficient transport of the drug into the bones, in which presented lowest concentration of Pt and, the highest concentration were found on the outside of the bone samples, confirming the hypothesis that the Pt drug did not penetrate the bones [55].

Studies involving NPs has increased in the last years because of their versatility of applications, such as in medical science (drug delivery, label method or immunotherapy) and environmental (nutrient or toxic transport) [55–57]. Imaging the distribution of Ag, Cu, and Mn in soybean leaves cultivated in the presence or the absence of AgNPs (40 nm average size) and also using $AgNO_3$ indicated Ag poor translocation to leaves. Additionally, the homeostasis of Mn and Cu is highly affected in plants cultivated in the presence of $AgNO_3$ compared to those cultivated in the presence of AgNPs (**Figure 3**) [56]. Additionally, the study of bio-distribution of NPs is a challenge because of their small size (< 100 nm), so high spatial resolution imaging techniques present as promising assessment tools to evaluation at the cellular level and macroscales.

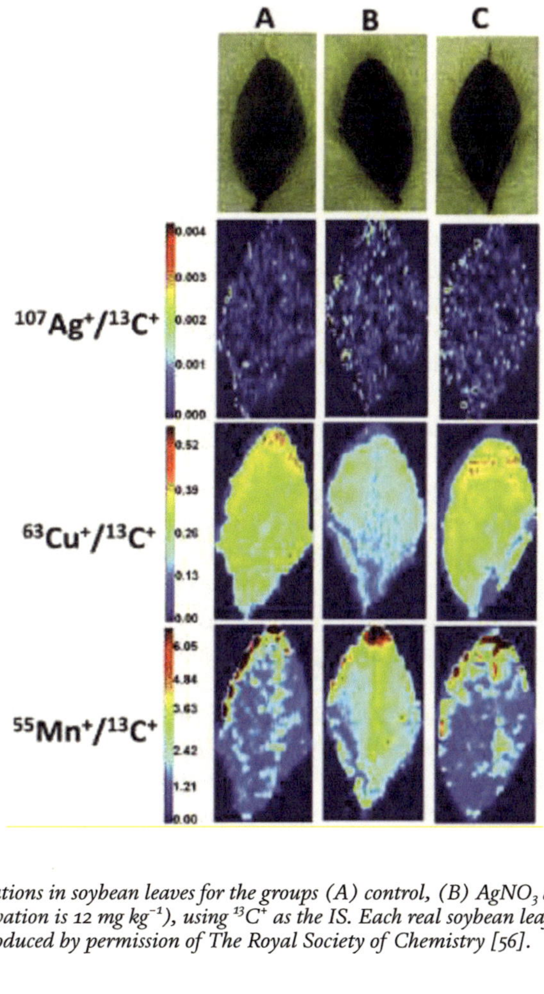

Figure 3.
Ag, Cu and Mn distributions in soybean leaves for the groups (A) control, (B) $AgNO_3$ and (C) AgNPs (total silver at the end of cultivation is 12 mg kg^{-1}), using $^{13}C^+$ as the IS. Each real soybean leaf is presented on the top of this figure [56] - reproduced by permission of The Royal Society of Chemistry [56].

Krajcarová *et al.* used double-pulse LIBS image mapping to verify the spatial distribution of Ag^+ and AgNPs in a small root cross-section of *Vicia faba* and their differences in transport efficiency. Ag^+ ions are toxic for plant tissues and the AgNPs are the most used as an antimicrobial agent in many products applied in plants. However, the later can interfere in the growth and biomass production if absorbed by the roots of the plants. Using a 50 μm of image resolution, the authors observed a clear difference between the localization of Ag^+ ions and AgNPs. AgNPs were found in just one part of root layers, close to the rhizodermis, and had no significant visual effects on the roots, just reduced the lateral root formation. Meanwhile, Ag^+ presented higher signal intensities and homogeneous distribution in the root cortex and caused reduced root lengths, darker color, and absence of lateral roots [58].

LA-ICP-MS imaging and NPs was used by Cruz-Alonso *et al.* to evaluate molecular distribution and quantification in human retina sections of 4 postmortem donors. Gold nanoparticles were applied as antibody-conjugated nanoclusters (Ab-AuNCs) to identify metallothioneins (MTs), which are protective to the neural retinal cells against oxidative stress. The LA-ICP-MS quantitative imaging (expressed in ng of MT g^{-1}) was applied in line scan mode with 10 μm of spot diameter and a matrix-matched analytical curve was used as a calibration strategy. As the sample is a biological tissue, the authors replaced the commercial ablation chamber by an in-house chamber with a reduced internal volume and cell temperature constant at −20°C, keeping the sample integrity. As results, the Ab-AuNCs allowed observed that the MTs present distribution in some principals retinas layers, known inner and outer nuclear layers and ganglion cells layers, as well as a present different concentration between the 4 postmortem donors, which is correlated with the biological diversity characteristics of each individual patient [52].

Different applications of imaging process applied to environmental studies [22], and translocation and accumulation of metals in plant tissues [59]. Paleoclimate and geological researches developed important studies in environmental science aiming to understand climate and environmental impacts using tree rings and speleothems as natural archive samples. These samples present high temporal resolution, allowing temporal reconstruction because changes in environmental or climate conditions can interfere on growth periods and elemental composition. Locosseli *et al.* demonstrated with LA-ICP-MS imaging that the decrease of Pb through the tree rings of *Tipuana tipu* (Fabaceae) could be correlated to the decrease of Pb concentration in gasoline of São Paulo (Brazil). In this study, high-resolution scan mode with x-resolution of 31.5 μm, scan speed varying of 60–100 μm s^{-1}, and LA-iMageS software was used [60].

A limitation in environmental studies is the small sample size that is allowed in the ablation chamber used in LA-ICP-MS, limiting experimental sampling. As an advantage of LIBS imaging, the ablation chamber does not be needed to be closed. Recently, Cáceres *et al.* reported the development of a method that enables the create megapixels elemental image of large speleothems (25 cm long) and coral (8 cm long) surfaces by fast and high-resolution LIBS imaging. Approximately 360,000 pixels $hour^{-1}$ were obtained in a 10 μm of lateral resolution, 100 Hz (100 pixels s^{-1}) as operation speed, and 8 hours per 2D imaging created with LasMap software. It allowed the visualization of fibers and specific regions of coral with the Mg/Ca, Sr/Ca and Na/Ca images, and according to the authors, the speleothems images demonstrated that compositional variations of Sr/Ca and Mg/Ca are not random but they result of variations of paleoenvironmental proxies [61].

As previously mentioned, LA-ICP-MS and LIBS can have the same initial laser system (**Figure 1**), which is a multimodal method in the imaging process.

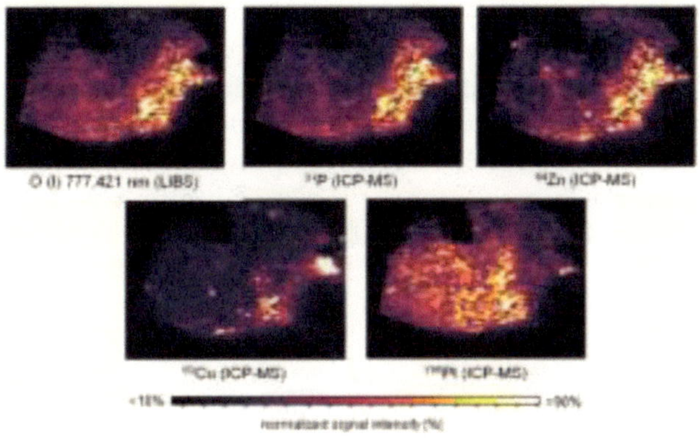

Figure 4.
Multimodal images of human tumor. Adapted from Bonta et al. [62] with permission from The Royal Society of Chemistry.

This approach was applied for Bonta *et al.* to create an elemental mapping of biological tissues with a tumor of human malignant pleural mesothelioma (MPM) (**Figure 4**).

Using a line scan with 40 μm of spot diameter, 40 μm of lateral resolution, 80 μm s^{-1} of scan speed and the software ImageLab to create the images, the authors observed that O, P, Zn, and Cu exhibit similar heterogeneous spatial distribution in the tissue, with high signal intensity in the tumor region, correlated with high number of activated proteins and their cofactors in this area. Different elements were detected in the tumor area, the presence of Pt, originated from common anti-cancer drug cisplatin, was detected in the healthy area, but not in the tumor's region [62]. This information would not be observed using a conventional method of analysis via decomposition.

5. Conclusions

Remarkable research and developments in laser technology and atomic spectrometry have promoted LIBS and LA-ICP-MS techniques to the height of its maturity in terms of both instrumentation and applications. Sensitivity, precision, and accuracy have continuously been improved, and obviously dependent on the experimental configurations used (type of laser, spectrometers, and detectors). These techniques allow the direct analysis of solids which is a significant advantage for the analysis of difficult-to-digest materials. High spatial resolution capacity allows for detailed information of the sample surface composition. All these features result in promising solutions for studies which the spatially resolved elemental information are required to provide a better understanding of the sample with significant advantages over conventional bulk analyses, such as in medical, environmental and technological science. Because LIBS present simpler instrumentation and an interesting trend is the portable instruments particularly attractive for in situ applications such as in agriculture and environmental analysis. Still, some researches focus on quantitative developments and calibration strategies to overcome remaining drawbacks such as the lack of CRMs, matrix effects and fractionation as well as multimodal system and isotopic analysis by LA-ICP-MS.

At this stage of development, we may point out that LA-ICP-MS and LIBS have complementary analytical performances and may be considered as attractive alternatives for the analysis of both bulk samples and elemental imaging distribution.

Acknowledgements

The authors thank the São Paulo Research Foundation (FAPESP, 2018/25207-0, 2017/50085-3, 2018/23478-7, 2019/00063-9, 2019/00018-3), the Brazilian National Council of Scientific and Technological Development (CNPq, 401170/2016-0), the Coordination for the Improvement of Higher Education Personnel (CAPES, 88887.115406/2015) and INCTBio (FAPESP, 2014/50867-3) for financial support and fellowships.

Author details

Renata S. Amais, Danielle S. Francischini, Pedro S. Moreau and Marco A.Z. Arruda*
Spectrometry, Sample Preparation and Mechanization Group - GEPAM, Institute of Chemistry, University of Campinas, UNICAMP, Campinas, Brazil

*Address all correspondence to: zezzi@unicamp.br

References

[1] O'Connor PB, Hillenkamp F. MALDI Mass Spectrometry Instrumentation, In: Hillenkamp F, Peter-Katalinić J, editors. MALDI MS: A Practical Guide to Instrumentation, Methods and Applications. Chichester: John Wiley & Sons, Inc.; 2007. 29-82 p. DOI: 10.1002/9783527610464

[2] Horn I. Laser Ablation in Atomic Spectroscopy, In: Meyers RA, editors. Encyclopedia of Analytical Chemistry. Chichester: John Wiley & Sons, Inc.; 2010. 1-19 p. DOI: 10.1002/9780470027318.a5110m.pub2

[3] Kompa KL, Laser Sources for Chemical Experiments. In: Kompa KL, Wanner J, editors. Laser Applications in Chemistry. Boston: Springer; 1984; 105. 15-24 p. 10.1007/978-1-4613-2739-4_2

[4] Brisbane A. Syracuse Post Standard, 1911; 28: 18.

[5] Oliveira SR, Arruda MAZ. Application of laser ablation (imaging) inductively coupled plasma mass spectrometry for mapping and quantifying Fe in transgenic and non-transgenic soybean leaves. J. Anal. At. Spectrom. 2015;30:389-395. DOI: 10.1039/C4JA00276H.

[6] Brech F, Cross L. Optical microemission simulated by a ruby maser. Appl. Spectrosc. 1962;16:59-64.

[7] Baudelet M, Smith BW. The first years of laser-induced breakdown spectroscopy. J. Anal. At. Spectrom. 2013;28:624-629. DOI: 10.1039/c3ja50027f

[8] Cremers DA, Multari RA, Knight AK. Laser-induced Breakdown Spectroscopy, In: Meyers RA, editors. Encyclopedia of Analytical Chemistry. Chichester: John Wiley & Sons, Inc.; 2010. 1-28 p. DOI: 10.1002/9780470027318.a5110t.pub3

[9] Pasquini C, Cortez J, Silva LMC, Gonzaga FB. Laser Induced Breakdown Spectroscopy. J. Braz. Chem. Soc. 2007;18:463-512. DOI: 10.1590/S0103-50532007000300002

[10] Anabitarte F, Cobo A, Lopez-Higuera JM. Laser-Induced Breakdown Spectroscopy: Fundamentals, Applications, and Challenges. ISRN Spectroscopy. 2012;2012:285240. DOI:10.5402/2012/285240

[11] Zhang S, Zhang B, Hang W, Huang B. Chemometrics and theoretical approaches for evaluation of matrix effect in laser ablation and ionization of metal samples. Spectrochimica Acta - Part B: At Spectroscopy. 2015;107:17-24. DOI:10.1016/j.sab.2015.02.009.

[12] Cremers DA, Radziemski LJ. Handbook of Laser-Induced Breakdown Spectroscopy. Chichester: John Wiley & Sons, Inc.; 2006. 23-52 p. DOI: 10.1002/0470093013

[13] Hahn DW, OMENETTO N. Laser-Induced Breakdown Spectroscopy (LIBS), Part II: Review of Instrumental and Methodological Approaches to Material Analysis and Applications to Different Fields. App. Spectrosc. 2012;66:347-419. DOI: 10.1366/11-06574

[14] Zhang S, He M, Yin Z, Zhu E, Hang W, Huang B. Elemental fractionation and matrix effects in laser sampling based spectrometry. J. Anal. At. Spectrom. 2016,31:358-382. DOI:10.1039/C5JA00273G

[15] Bulajic D, Corsi M, Cristoforetti G, Legnaioli S, Palleschi V, Salvetti A, Tognoni E. A procedure for correcting self-absorption in calibration free-laser induced breakdown spectroscopy. Spectrochim. Acta Part B. 2002;57:339-353. DOI: 10.1016/S0584-8547(01)00398-6

[16] Gray A L. Solid sample introduction by laser ablation for inductively coupled plasma source mass spectrometry. Analyst. 1985;110:551-556. DOI: 10.1039/AN9851000551

[17] Cremers DA, Radziemski LJ. Handbook of Laser-Induced Breakdown Spectroscopy. Chichester: John Wiley & Sons, Inc.; 2006. 53-98 p. DOI: 10.1002/0470093013

[18] Fu X, Li G, Dong D, Improving the Detection Sensitivity for Laser-Induced Breakdown Spectroscopy: A Review. Front. Phys. 2020;8:68. DOI: 10.3389/fphy.2020.00068

[19] Kim C-K, In J-H, Lee S-H, Jeong S. Influence of Ar buffer gas on the LIBS signal of thin CIGS films. J. Anal. At. Spectrom. 2013;28:460-467. DOI: 10.1039/C2JA30297G

[20] Gaft M, Nagli L, Eliezer N, Groisman Y., Forn O. Elemental analysis of halogens using molecular emission bylaser-induced breakdown spectroscopy in air. Spectrochim. Acta Part B. 2014;98:39-47. DOI: 10.1016/j.sab.2014.05.011

[21] Resano M, Aramendía M, Nakadi FV, García-Ruiz E, Alvarez-Llamas C, Bordeld N, Pisonero J, Bolea-Fernández E, Liu T, Vanhaecke F. Breaking the boundaries in spectrometry. Molecular analysis with atomic spectrometric techniques. TRAC - Trends Anal. Chem. 2020;129:1159552. DOI: 10.1016/j.trac.2020.115955

[22] Jolivet L, Leprince M, Moncayo S, Sorbier L, Lienemann C.-P., Motto-Ros V. Review of the recent advances and applications of LIBS-based imaging. Spectrochim. Acta Part B. 2019;151:41-53. DOI: 10.1016/j.sab.2018.11.008

[23] Russo RE, Mao X, Gonzalez JJ, Zorba V, Yoo J. Laser ablation in analytical chemistry. Analytical Chemistry. 2013;85:6162-6177. DOI:10.1021/ac4005327

[24] Russo RE, Mao X, Liu H, Gonzalez J, Mao SS. Laser ablation in analytical chemistry-a review. Talanta. 2002;57:425-51. DOI:10,1016/S0039-9140(02)00053-X

[25] Russo RE, Mao X, Borisov O V., Liu H. Laser Ablation in Atomic Spectroscopy. Encyclopedia of Analytical Chemistry. 2006;1-22. DOI: 10.1002/9780470027318.a5110m

[26] Gurevich EL, Hergenröder R. A simple laser ICP-MS ablation cell with wash-out time less than 100 ms. J. Anal. At. Spectrom. 2007;22:1043-1050. DOI: 10.1039/b704700b

[27] Frick DA, Günther D. Fundamental studies on the ablation behaviour of carbon in LA-ICP-MS with respect to the suitability as internal standard. Journal of Analytical Atomic Spectrometry. 2012;27(8):1294-1303. DOI:10.1039/c2ja30072a

[28] Van Malderen SJM, Managh AJ, Sharp BL, Vanhaecke F. Recent developments in the design of rapid response cells for laser ablation-inductively coupled plasma-mass spectrometry and their impact on bioimaging applications. J. Anal. At. Spectrom. 2016;31:423-439. DOI: 10.1039/C5JA00430F

[29] Evans EH. Atomic Mass Spectrometry | Inductively Coupled Plasma. Encyclopedia Analytical Science (Second Edition) 2005;229-37. DOI:10.1016/B0-12-369397-7/00036-4

[30] Thomas R. Practical guide to ICP-MS: a tutorial for beginners. Third. CRC Press Taylos & Francis Group; 2013. 409 p. ISBN:9781466555433

[31] Miliszkiewicz N, Walas S, Tobiasz A. Current approaches to calibration of LA-ICP-MS analysis. J. Anal. At.

Spectrom. 2015;30(2):327-38. DOI: 10.1039/c4ja00325j

[32] Kuczelinis F, Weis P, Bings N-H. Forensic comparison of PVC tape backings using time resolved LA-ICP-MS analysis. Forensic Chem. 2019;12:33-41. DOI:10.1016/j.forc.2018.11.004

[33] Corzo R, Hoffman T, Weis P, Franco-Pedroso J, Ramos D, Almirall J. The use of LA-ICP-MS databases to calculate likelihood ratios for the forensic analysis of glass evidence. Talanta. 2018;186:655-61. DOI:10.1016/j.talanta.2018.02.027

[34] Ash RD, He M. Details of a thallium poisoning case revealed by single hair analysis using laser ablation inductively coupled plasma mass spectrometry. Forensic Sci. Int. 2018;292:224-231. DOI: 10.1016/j.forsciint.2018.10.002

[35] Noël M, Christensen JR, Spence J, Robbins CT. Using laser ablation inductively coupled plasma mass spectrometry (LA-ICP-MS) to characterize copper, zinc and mercury along grizzly bear hair providing estimate of diet. Sci. Total Environ. 2015;529:1-9. DOI:10.1016/j.scitotenv.2015.05.004

[36] Yan N, Zhu Z, Jin L, Guo W, Gan Y, Hu S. Quantitative Characterization of Gold Nanoparticles by Coupling Thin Layer Chromatography with Laser Ablation Inductively Coupled Plasma Mass Spectrometry. Anal. Chem. 2015;87(12):6079-6087. DOI:10.1021/acs.analchem.5b00612

[37] Vorapalawut N, Labrador MM, Pohl P, Caetano M, Chirinos J, Arnaudguilhem C, et al. Application of TLC and LA ICP SF MS for speciation of S, Ni and V in petroleum samples. Talanta. 2012;97:574-578. DOI:10.1016/j.talanta.2012.04.040

[38] Lobo L, Pereiro R, Fernández B. Opportunities and challenges of isotopic analysis by laser ablation ICP-MS in biological studies. Vol. 105, TrAC - Trend Anal. Chem. 2018;105:380-390. DOI:10.1016/j.trac.2018.05.020

[39] Limbeck A, Galled P, Bonta M, Bauer G, Nischkauer W, Vanhaecke F. Recent advances in quantitative LA-ICP-MS analysis: challenges and solutions in the life sciences and environmental chemistry. Anal. Bioanal. Chem. 2015;407:6593-6617. DOI:10.1007/s00216-015-8858-0

[40] Pozebon D, Scheffler GL, Dressler VL. Recent applications of laser ablation inductively coupled plasma mass spectrometry (LA-ICP-MS) for biological sample analysis: A follow-up review. J. Anal. At. Spectrom., 2017;32(5):890-919. DOI:10.1039/c7ja00026j

[41] Payne TD, Moody AS, Wood AL, et al. Raman spectroscopy and neuroscience: From fundamental understanding to disease diagnostics and imaging. Analyst. 2020;145:3461-3480. DOI: 10.1039/d0an00083c

[42] Grubman A, James SA, James J, et al. X-ray fluorescence imaging reveals subcellular biometal disturbances in a childhood neurodegenerative disorder. Chem. Sci. 2014; 5: 2503-2516. DOI: 10.1039/C4SC00316K

[43] Nuñez J, Renslow R, Cliff JB, et al. NanoSIMS for biological applications: Current practices and analyses. Biointerphases 2018; 13: 03B301. DOI: 10.1116/1.499362

[44] Koch J, Günter D. Laser Ablation Inductively Coupled Plasma Mass Spectrometry. Encycl. Spectrosc. Spectrom. 2016; 526-532. DOI: 10.1016/B978-0-12-803224-4.00024-8

[45] Castellanos-García LJ, Gokhan Elci S, Vachet RW. Reconstruction,

analysis, and segmentation of LA-ICP-MS imaging data using Python for the identification of sub-organ regions in tissues. Analyst. 2020;145:3705-3712. DOI: 10.1039/c9an02472g

[46] Gimenez Y, Busser B, Trichard F, et al. 3D Imaging of Nanoparticle Distribution in Biological Tissue by Laser-Induced Breakdown Spectroscopy. Sci. Rep. 2016;6:29936. DOI: 10.1038/srep29936

[47] Fernández HL, Pessôa GDS, Arruda MAZ, et al. LA - iMageS : a software for elemental distribution bioimaging using LA – ICP – MS data. J. Cheminform. 2016;1-10. DOI: 10.1186/s13321-016-0178-7

[48] Pessôa GDS, Capelo-Martínez JL, Fdez-Riverola F, et al. Laser ablation and inductively coupled plasma mass spectrometry focusing on bioimaging from elemental distribution using MatLab software: A practical guide. J. Anal. At. Spectrom. 2016;31:832-840. DOI: 10.1039/c5ja00451a

[49] Pořízka P, Klus J, Képeš E, et al. On the utilization of principal components analysis in laser-induced breakdown spectroscopy data analysis, a review. Spectrochim. Acta Part B. 2018;148:65-82. DOI: 10.1016/j.sab.2018.05.030

[50] Modlitbová P, Pořízka P, Kaiser J. Laser-induced breakdown spectroscopy as a promising tool in the elemental bioimaging of plant tissues. TrAC - Trends Anal. Chem. 2019;122:115729. DOI: 10.1016/j.trac.2019.115729

[51] Westerhausen MT, Bishop DP, Dowd A, et al. Super-Resolution Reconstruction for Two-and Three-Dimensional LA-ICP-MS Bioimaging. Anal. Chem. 2019;91:14879-14886. DOI: 10.1021/acs.analchem.9b02380

[52] Cruz-Alonso M, Fernandez B, García M, et al. Quantitative Imaging of Specific Proteins in the Human Retina by Laser Ablation ICPMS using Bioconjugated Metal Nanoclusters as Labels′. Anal. Chem. 2018;90:12145-12151. DOI: 10.1021/acs.analchem.8b03124

[53] Sussulini A, Becker JS, Becker. Laser ablation ICP-MS: application in biomedical research. Mass Spectrom. Rev. 2017;36:47-57. DOI: 10.1002/mas.21481

[54] Moncayo S, Trichard F, Busser B, et al. Multi-elemental imaging of paraffin-embedded human samples by laser-induced breakdown spectroscopy. Spectrochim. Acta - Part B. 2017;133:40-44. DOI: 10.1016/j.sab.2017.04.013

[55] Crone B, Schlatt L, Nadar RA, et al. Quantitative imaging of platinum-based antitumor complexes in bone tissue samples using LA-ICP-MS. J. Trace Elem. Med. Biol. 2019;54:98-102. DOI: 10.1016/j.jtemb.2019.04.011

[56] Chacón-Madrid K, Arruda MAZ. Internal standard evaluation for bioimaging soybean leaves through laser ablation inductively coupled plasma mass spectrometry: a plant nanotechnology approach. J. Anal. At. Spectrom. 2018;33:1720-1728. DOI: 10.1039/c8ja00254a

[57] Le Guével X, Henry M, Motto-Ros V, et al. Elemental and optical imaging evaluation of zwitterionic gold nanoclusters in glioblastoma mouse models. Nanoscale. 2018;10:18657-18664. DOI: 10.1039/C8NR05299A

[58] Krajcarová L, Novotný K, Kummerová M, et al. Mapping of the spatial distribution of silver nanoparticles in root tissues of *Vicia faba* by laser-induced breakdown spectroscopy (LIBS). Talanta. 2017;173:28-35. DOI: 10.1016/j.talanta.2017.05.055

[59] Pessôa GDS, Lopes CA, Madrid KC, et al. A quantitative approach for Cd , Cu , Fe and Mn through laser ablation imaging for evaluating the translocation and accumulation of metals in sun flower seeds. Talanta. 2017;167:317-324. DOI: 10.1016/j.talanta.2017.02.029

[60] Locosselli GM, Chaccon-Madrid K, Arruda MAZA, et al. Tree rings reveal the reduction of Cd , Cu , Ni and Pb pollution in the central region of São Paulo, Brazil. Environ. Pollut. 2018;242:320-328. DOI: 10.1016/j.envpol.2018.06.098

[61] Cáceres JO, Pelascini F, Moncayo S, et al. Megapixel multi-elemental imaging by Laser-Induced Breakdown Spectroscopy, a technology with considerable potential for paleoclimate studies. Sci. Rep. 2017;7:5080. DOI: 10.1038/s41598-017-05437-3

[62] Bonta M, Gonzalez JJ, Quarles Jr. CD, et al. Elemental mapping of biological samples by the combined use of LIBS and LA-ICP-MS. J. Anal. At. Spectrom. 2016;31:252-258. DOI: 10.1039/C5JA00287G